THE YOUNG MAN'S

BEST COMPANION,

OR

MATHEMATICAL COMPENDUM,

Containing a great variety of very useful

RULES AND EXAMPLES IN MATHEMATICS

FOR THE

Merchant, Clerk, Accountant and the Mechanic,

Worked out so as to be
QUICKLY UNDERSTOOD BY ANY ONE
WHO UNDERSTANDS THE

FOUR FUNDAMENTAL RULES OF ARITHMETIC

BY

AMOS W. WARREN.

RUTLAND:
TUTTLE & CO., PRINTERS,
1872.

PREFACE.

It has been said by one wiser than we, "despise not the day of small things."

The want of a book like the Young Man's Best Companion, in which the principles of mathematics are clearly demonstrated, has long been felt. Others of a larger size for schools and academies have been multiplied.

The young man will herein find a great variety of very useful rules and examples worked out so as to be easily understood; suited to ordinary business so that but little time is required to refresh the mind with the desired information, for an author says "if they do not daily practice the majority of people forget."

Let the young man, the clerk, merchant, accountant, mechanic, or the proprietor of the warehouse, have this readily at hand, and he is proof against any mistake incident to forgetfulness or any other cause.

This is the great secret of its peculiar adaptation to the business man. And the author hereby hopes that the public will realize in this little book what has so long been sought elsewhere in vain.

CONTENTS.

PERCENTAGE.

The term percentage and per cent. signifies a certain part of a hundred; thus, 4 out of a hundred, is 4 per cent.; 6 out of a hundred, is 6 per cent.; and so on for other rates per cent.

When we say 5 per cent of the population of the State of New York, we mean five persons out of every hundred.

RULE.—Find the amount by substraction, multiply the gain or loss by 100, and divide the product by the purchase price.

Ex. 1. The amount of my capital invested in business is $7,500, and I have sold goods to the amount of $8,400; what per cent is it?

```
                         900
                         100
    8400                -----
    7500         7500) 90000
    ----                -----
     900 gain.             12 per cent.  Ans.
```

Ex. 2. A hotel cost $350,000; was rented for $24,500 a year. What per cent did it rent for on the cost?

```
                           24500
                             100
                         -------
Purchase price, 350000) 2450000
                         -------
                               7 per cent.  Ans.
```

Ex. 3. A merchant received $862.50 dividend on $17,250 stock; what was the rate per cent.?

```
                862.50
                   100
             ---------
17250.00) 86250.00
             ---------
                      5 per cent.   Ans.
```

The $862.50 are hundredths; the dividend and diviser must be of the same denomination; we annex 2 ciphers to 17250 and it becomes 17250 00 hundredths.

Ex. 4. A merchant bought calico at 12 cts. per yard, and sold it at 12 1-2 cents per yard; what per cent did he gain? Ans. 4 1-6 per cent.

```
12.5*=12 1-2 cts.          100
12.  =12     "               5
---------------         ---------
   5 mills     120 mills) 500 mills.
                          ---------
                             4 1-6

12  cents per yard.
 10   "   a mill.
-----
120 mills.
```

Ex. 5. A man paid $24.54 insurance on $6544; what was the rate per cent.? Ans. $\frac{375}{1000}$

```
              24.54
                 100
            ---------
6544.00) 2454.000 (0.375
            1963200
            ---------
             4908000
             4580800
            ---------
              3272000
              3272000
```

*= signifies equality.

Annex ciphers to the diviser as Ex. 3, to make it of the same denomination as the dividend.

We now see that the decimal places in the dividend exceed those in the diviser by 3, counting the ciphers annexed making $\frac{375}{1000}$ per cent. the ans. See rule in Decimal Fractions.

Ex. 6. A bought goods amounting to $550; what per cent profit must he make to gain $66.

```
                        66   gain.
                        100
                        ----
Purchase price, 550)  6600
                        ----
                        12 per cent.   Ans.
```

Ex. 7. The Rutland Marble Co. sold Lyman Strong of Cleveland, O., a lot of marble for $300; being a wholesale dealer, 1-6 is discounted from his bill; what per cent. would S. make to sell it at its prime cost, exclusive of freight?

```
                                       5000
                                        100
                                     ------
                              25000) 500000
6) 30000                             ------
    5000                                 20  per ct.   Ans.
  ------
  250.00  purchase                    300.00
          price.                      250.00
                                      ------
                                       50.00 gain.
```

Ex 8. What is 8 per cent of $1567825.46?

```
                                     1567825.46
   8 per cent is written thus,              .08
                                   ------------
Ans. $125426.03.5                 $125426.03,68
```

There are four decimal places in both factors; we point in the product 4 decimal places; the figures at the left of the point will be dollars, and those at the right will be cents and mills, making one hundred and twenty-five thousand, four hundred and twenty-six dollars, three cents and six mills.

Ex. 9. What is 71 per cent of 8875794 bushels of wheat? Ans. $6301813\frac{74}{100}$ bus.

```
                                 8875794
71 per cent is written thus,         .71
                              ----------
                                 8875794
                                62130558
                              ----------
                              6301813.74
```

There are two decimal places in one factor; we point off two decimal places in the product; those at the left will be the whole number of bushels, and those at the right will be fractions of the same.

Ex. 10. What is 50 per cent of $875590? Ans. $437795.

```
1st method.                 2d method.
  875590.00                 2) 875590
        .50                    ------
-------------                  437795
$437795.0000
```

There are 4 decimals in both factors; we point off 4 decimals in the product; those at the left are dollars.

Ex. 11. What is 2 per cent of $350. Ans. $7.

```
                                 350.00
2 per cent is written thus,         .02
                               --------
                               $7,00.00
```

Point off in the product the same as in example No. 10.

Ex. 12. What is 3 per cent of $145.25. Ans. $4.35.

```
                                 145.25
3 per cent is written thus,         .03
                                -------
                                4.35.75
```

Point off in the product the same as example 8.

Ex. 8. What is 1-2 per cent of $180.42? Ans. 90c.

	1st method.	2d method.
	180.42	2)180.42
1-2 per cent is written thus,	.005	
		.90.21
In 1st method,	.90.210	

There are 5 decimal places in both factors; we point off 5 decimal places in the product; those at the right of the point will be cents and mills.

Ex. 14. A has a brick block of stores, which cost $16000, consisting of three stores, 4 offices and 3 tenements; what had it ought to rent for to make 10 per cent. Ans. $1600.

	16000.00
10 per cent is written thus.	.10
	$1600.0000

Point off in the product the same as Ex. 10.

Ex. 15. What is 3-4 per cent of $128.63? Ans. .96.4

	1st method.
	$128.63
3-4 per cent is written thus,	.0075
	64315
	90041
	.96.4.725

There are 6 decimal places in both factors; point off 6 decimal places in the product; all at the right of the point will be cents and mills.

2d method.

4.2) 128.63

6431=1-2

3215=3-4

.96.46

Ex. 16. A stockholder owned 10 shares of $100 each of the Hudson Railroad stock; received a dividend of $50 every 6 months; what per cent is it on the money invested?

```
$100  each share.
  10 shares.
-----
$1000.
                Recd. $100  a year.
                       100
                     -----
                1000) 10000
                     -----
                        10 per cent.  Ans.
```

Ex. 17. The population of New York in 1850 was 515,647, and in 1860 was 814,277; what per cent was the increase?

```
         814,277
         515,647
         -------
         298,630
             100
         ----------
515,547) 298,630.00
         ----------
               58 per cent., nearly.  Ans.
```

Ex. 18. A has 840 bbls. of flour and sold 20 per cent of it; how many bbls. did he sell. Ans. 168 bbls.

```
                                  840
20 per cent is written thus,      .20
                               ------
                               168.00
```

Point off in the product the same as Ex. 9.

Ex. 19. If a fleece of unwashed wool weighs 17 lbs.,

8 ozs., the same fleece weighs 7 lbs., 8 1-2 ozs., washed; what per cent of shrink in cleaning?

```
    17—8
    16 ozs. lb.
   ------
   280
   120.5                    159.5
   ------                     100
   159.5                   --------
                       280)15950.0(56.9
                           1400
 7—8 1-2                   ------
 16 ozs. lb.                1950
 ---------                  1680
 112                       ------
   8 1-2                     2700
 ---------                   2520
 120.5                       ----
```

The decimal places in the dividend exceed those in the divisor by one, hence we point off one decimal place in the quotient, making 56 9-10 per cent., the answer.

Ex. 20. The following report is the condition of the finances of the liabilities of the town of Rutland, Vermont:

Liabilities, - - - - -	$12,266 53
Resources, - - - - -	2,013 13
Balance of liabilities, - -	$10,253 40

They estimate the current expenses of the ensuing year as follows:

Building the Moulthrop road, -	$1,000	
Support of the Poor, -	1,300	
Other expenses, - - -	1,700	4,000
Amount to be paid for, - -		$14,253 40

The Grand List of the town is $24,456 28, and the amount to be provided for is $14,253 40; what per cent is required on the Grand List to meet the above tax.

```
            14253.40
                 100
            --------
24456.28)1425340.00(58.28 Ans.
          12228140
          --------
          20252600
          19565024
          --------
            6875760
            4891256
          --------
            198455040
            19565024
          --------
```

The decimal places in the dividend exceed those in the divisor by 2, counting the ciphers annexed, hence we point off in the quotient 2 places for hundreths, and those at the left hand will be the whole number of percentage, making $58\frac{28}{100}$ per cent.

PERCENTAGE TABLE.

1	per cent is written thus, - - - -	.01
2	" " " - - -	.02
3	" " " - - - -	.03
6	" " " - - -	.06
7	" " " - - - -	.07
10	" " " - - -	.10
12	" " " - - - .	.12
50	" " " - - -	.50
100	" " " - - - -	1.00
103	" " " - - -	1.03
125	" " " - - - -	1.25

1-2 per ct., that is,	1-2 1 per ct.,	written thus,	.005
1.4 " "	1-4 "	" "	.0025
3-4 " "	3-4 "	" "	.0075

100 per cent is $1.00\frac{100}{100}$ or the whole.

120 per cent is $1.20\frac{20}{100}$ more than the whole.

PERCENTAGE IN INTEREST.

To find the rate, per cent., when the principal and the interest and time is given.

RULE.—Divide the given interest by the interest of the principal at 1 per cent. for the given time, and the quotient will be the required per cent.

Ex. 1. A merchant borrowed $90 for 5 years, and paid $36 for the use of it; what is the rate per cent.?

No. 1. Ans. 8 per cent.

```
90.00                     90
  .01 per cent.            5 years.
 -----                   ----
.90.00=90 cts. per 1 year. 4.50 int. of the principal.
              4,50)36.00
                   -----
                     8 per cent.
```

There are 4 decimals in both factors of No. 1, hence we point off 4 decimals from the product; those at the right of the point will be cents.

Ex. 2. A merchant is worth $25,000, at what per cent. must he loan his capital that his income may be just $1,000.

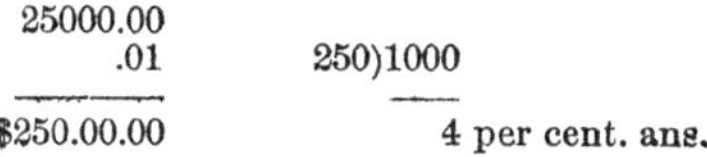

4 per cent. ans.

Point off the same as Ex. 10 in percentage.

INSURANCE.

Insurance is a contract by which one party engages, for a stipulated premium, to make a loss which another may sustain.

Rule.—Multiply the sum insured by the given rate per cent., expressed in decimals, and the product will be the premium. The same as interest and discount' Policy is the contract—premium is the tax on the property insured.

Ex. 1. A machine shop is insured for 5 years for $6,000; premium note 22 per cent., of which 4 1-2 per cent. was paid down, and 2 1-2 per cent. assessments; what did it cost per year? Ans. $18.48.

No. 1.	No. 2.
2)1320.00	6000.00
.02 1-2	22
264000	2)1320.0000
66000	.04 1.2
33.00.00	52800000
59.4000	6600000
5)92.40.00	59.40.0000
18.48	

Point off in Nos. 1 & 2 first products the same as Ex. 10 in percentage, and in 2d product of No. 2, 6 decimals; all at the left are dollars, and those at the right are cents.

Ex. 2. What is the premium on my house valued at $4365, at 1-2 per cent.? Ans. $21.82.5.

```
                                    4365.00
1-2 per cent. is written thus,         .005
                                 ----------
                                  21.82.500
```

We point off 5 decimals in the product; all at the left of the point are dollars, those at the right are cents and mills.

What is the insurance on $15550 at 1 1-2 per cent.? Ans. 233.25.

```
                                   15550.00
1 1-2 per cent. is written thus,       .015
                                 ----------
                                 233.25 000
```

Point off in the product the same as Ex. 2.

Ex. 4. If a policy be taken out of $6400 at 4 per cent., what net amount is it after paying the insurance?

```
 6400          6400.00
  256             .04=4 per ct.
-----       ----------
$6144 Ans.  $256 00.00
```

There are four decimals in the two factors; point off in the product the same as Ex. 2, Percentage in Interest.

Ex. 5. If the premium is 5 per cent., for what amount must a policy be taken out to cover $8835, together with the premium paid for insurance?

Ans. $9300.

```
                100 per ct. is written thus, 1.00
Deduct            5  "     "     "            .05
5 per ct.       ---------------------------------
makes            95  "     "     "            .95)8835.00(9300
                                                  855
                                                  ---
                                                  285
                                                  285
                                                  ---
                                                   00
```

When the decimal places in the divisor exceed those in the dividend, make them equal by annexing ciphers to the right of the dividend, and the quotient will be a whole number.

Ex. 6. A woolen factory and contents valued at $64800, and insured at 2 4-5 per cent., if destroyed by fire, what would be the actual loss of the company?

```
64800                  64800 00
 1814.40                  .028=2 4-5
--------               -----------
$62985.60 Ans.          51840000
                       12960000
                      ----------
                     $1814.40.000
```

There are 5 decimals in both factors; point off the same as Ex. 2.

Ex. 7. What is the premium on my household furniture valued at $500, premium note 8 per cent.; paid 3 per cent. on premium note, $2.10 for policy and application. Ans. $3.30.

```
   500.00
      .08
   ------
  40.0000
      .03
  -------
 1.20.0000
 2.10
 ----
$3.30
```

There are four decimals in the first two factors, hence we point off in the product the same as Ex. 10 in Percentage. There are 6 decimals in the next two factors; we point off the same as Ex. 1 in No. 2, 2d product.

STOCKS.

By the term stocks is meant the capital of monied institutions, incorporated banks, manufactories, railroads, insurance companies and State bonds. The original cost of a share is called its nominal or par* value. In most companies $100 is called a share, though in some companies more and in some less, but the market value varies according to circumstances.

RULE.—Multiply the par value of the stock by the rate per cent., decimally.

Ex. 1. A buys 20 shares of Boston & Worcester R. R. stock at 7 per cent. advance; how much did his stock cost, original shares $100. Ans. $2140.

100 per cent. is written thus, 1.00
7 " " " " .07
107 " " " " 1.07

Add 7 per ct. makes

2000.00
1.07
———
14000.00
200000
———
2140.0000

Point off in the product the same as Ex. 10 in Percentage.

Ex. 2. What will 20 shares of the same stock amount to at 7 per cent. discount? Ans. $1860.

*By par value is meant the original cost or estimate value of stock. When it is worth more than the original cost it is said to be above par. When it is worth less than the original cost it is said to be below par.

100 per cent. is written thus, 1.00
7 " " " " .07

93 " " " " .93

Deduct 7 per ct. makes

2000.00
.93

600000
1800000

1860.0000

Point off in the product the same as Ex. 10.

Ex. 3. A merchant employed a broker to buy 55 shares of bank stock, which is 20 per cent. below par, and pays him 3-4 per cent. brokerage; how much will his stock cost?

100 per cent. is written thus, 1.00
20 " " " " .20

80 " " " " 80

Deducting 20 per ct. makes

$100.00 a share
55 shares

5500.00
.80

4400.0000
41.25

4441.25 ans.

4.2)5500

2750=1-2
1375=3-4

$41.25 and brokerage.

Ex. 4. A stock jobber bought 45 shares of the Vt. Central R. R. stock at 3 per cent. discount, which he sold at 7 per cent. advance; how much did he make, the original shares $100.

100 per cent. is written thus, 1.00
3 " " " " .03

.97 " " " " .97

Deduct 3 per ct. makes

100.00
45

4500.00
.97

4365.0000

```
     100 per cent. is written thus, 1.00
       7   "    "    "    "    .07
     ---------------------------------
     1·07  "    "    "    "    1.07
Add                   4500.00
7 per ct.                1.07
makes                 -------
                    4815.0000
                    4365.0000
                    ---------
                    $450 ans.
```

Point off in the product the same as above.

Ex. 5. A cotton factory valued at $12000, is divided into 100 shares; if the profits amount to 15 per cent. yearly, what will be the profits accruing to 25 shares?

```
100 shares)12000                     Ans. $450
           -----
           $120 a share    120.00 a share
                               25 sharss
                           ------
                          3000.00
15 per cent. is written thus,  .15
                        ---------
                        $450.00.00
```

There are 2 decimals in one of the factors; we point off 2 decimals in the product; all at the left are dollars. In the 2d product the same as Ex. 10 in Percentage.

In the above factory, repairs are to be made which cost $340, what will be the tax on 10 shares?

```
100 shares)340
           ----
          $3.40 a share
             10 shares
          -----
         $34.00 ans.
```

Ex. 6. A farmer bought 52 shares of the Missisquoi

Bank and sold it at 13 per cent. premium ; how much did the stock come to ? Ans. $5876.

```
                              5200.00 am't of shares
113 per cent is written thus,    1.13
                              --------
                              $5876.0000
```

Point off in the product the same as above.

Ex. 7. Cornelius Vanderbilt owns 17000 shares of the Hudson River Railroad ; what amount has he invested ?

```
    17000 shares
        100=$100 a share
    ----------
$1,700,000
```

It reads one million, seven hundred thousand dollars, the answer.

Ex. 8. If the Rutland & Burlington Railroad declare an annual dividend of 12 per cent., what will a stockholder receive who owns 250 shares ?

```
                                    250       Ans. $3000
                                 100.00
                              ---------
                               25000.00
12 per cent. is written thus,       .12
                              ---------
                             $300.00.000
```

Point off in the product in the same manner as above.

Ex. 9. If the capital stock of a bank be $600,000, what amount is necessary to declare a dividend of 8 per cent. Ans. $48000

```
                              600000.00
8 per cent is written thus,         .08
                              ---------
                              48000.0000
```

Point off in the product the same as Ex. 10.

Ex. 10. What is the value of 10 shares in the Rutland Gas Light Co. at 85 per cent.? Ans. $850

```
                               100.00  amt. of share
                                   10
                              -------
                              1000 00
85 per cent. is written thus,     .85  per cent.
                             --------
                             850,0000
```

Point off in the same manner as above.

Ex. 11. A bank went into operation with a capital of $100,000, it is divided into 1000 shares of $100 each; at the end of 6 months the profit amounted to $5000; what per cent is it on the capital? what will be the profit on one share?

Ans. to the first, 5 per cent.
" " last, $5 on a share.

```
        5000
         100
       ------            100.00 a share
100000)500000               .05
       ------            ------
         .05 per cent.   $5.00.00
```

LOSS AND GAIN.

Loss and gain is an excellent rule by which merchants discover their profit or loss per cent. It also instructs them to raise or fall the price of their goods so as to gain or lose so much per cent., &c.

Rule.—Find the gain or loss, as the case may be, by substraction, then annex two ciphers, or multiply it by 100, and divide it by the first cost, the quotient will be the per cent.

Ex. 1. A bought broadcloth at $4.50 per yard, and sold the same for $5.50 per yard; what was the gain per cent?

```
5.50 price sold for.          100
4.50 first cost.                100
----                          -----
1.00                      450)10000
                              -----
                              22 2-9 per cent., ans
```

Ex. 2. Bought broadcloth at $4.00 per yard and sold the same at $3.50 per yard; what was the per cent. lost?

```
400
350              400)5000     2 ciphers annexed.
---                  ----
 50                  12 1-2 per cent., ans.
```

Ex. 3. Bought 63 gallons of molasses at 42 cents per gallon, but by accident, 8 gallons leaked out; how shall the remainder be sold, per gallon, to gain upon cost, at the rate of 10 per cent?

Ans., 52 cents, 9 mills.

```
100 per cent. is written thus, 1.00        63 gals.
 10   "     "    "     "        .10        42 cts.
---------------------------------------    -----
110   "     "    "     "       1.10        26.46
Add 10 per ct. makes                        1.10
          63 gals. bought.                 -----
           8  "   leaked out.          55)29.1060(0.52.9.2
          --                              275
          55  "   remaining.              ---
                                           160
                                           110
                                           ---
                                           506
                                           495
                                           ---
                                           110
                                           110
                                           ---
```

We point off in the first product 2 decimals; those at the left will be dollars; those at the right will be cents, and in the 2d product, the same as Ex. 8 page 3. We now see that the decimal places in the dividend exceed those in the divisor by 4, which we point off in the quotient; those at the right hand will be cents and mills.

Ex. 4. Bought 16500 gals. of oil for $8000, allowing 1 1-2 per cent. leakage; how much must it be sold for to gain 15 per cent.?

100 per cent. is written thus, 1.00
15 " " " " .15

115 " and " " 1.15

Adding 15 per ct. makes

165 0
.015=1 1-2 per cent.

247.500

There are 3 decimals in one factor; we point off from the product 3 decimal places; the figures at the left are the whole number of gallons, and those at the right are fractions of the same.

8000
1.15

16253)9200.00

056.6

16500
247

16253 gals. remaining.

Point off from the product or dividend the same as Ex. 3, 1st product, page 18.

The decimal places in the dividend exceed those in the divisor by 3, counting the cipher annexed, hence we point off in the quotient 3 decimals, making 56 cts. 6 mills.

Ex. 5. Bought 480 yds. linen at 82 cts. per yard, which shrank in bleaching 1 1-2 per cent.; after keeping it 6 mos. it was sold on 4 mos. credit at 20 per cent. advance on a yard; what was made, allowing 6 per cent. on the money invested?

Ans., 98 cts. 4 mills per yard,
and made $51.48.2.

```
100 per cent. is written thus, 1.00
 20    "     "     "     "     .20
-----------------------------------
120    "    and    "     "    1.20
Adding 20 per ct. makes
            No. 1.                        No.2.
           480 yards.                    472.32
           .82 cents per yard.             .984
          ------                     ----------
          393.60                     464.76.288
            1.20      No. 3.         413.28
          ------      393.60         ------
   480)472.3200          50=1-2 the mos.  51.48.2
       -------        ------
        0.98.40       19.68.8000
                      393.60
                      ------
                      413.28
```

Point off from the 1st product of No. 1 the same as Ex. 3, 1st product, page 18, and in 2d product or dividend, 4 decimals; those at the left will be dollars, and those at the right will be cents.

The decimal places in the product or dividend exceed those in the divisor by 4; hence we point off 4 decimals from the quotient.

In the product of No. 2, the same number decimals as Ex. 2, page 11, and in the product of No. 3, according to the rule in Simple Interest.

Ex. 6. Bought a house for $4280; at what price must it be sold so as to gain 30 per cent.?

```
                                 4280.00
130 per cent is written thus,       1.30
                                 -------
                               5564.0000
```

Point off in the product the same as Ex. 10, in percentage, page 5.

Ex. 7. Bought goods for $7500 and retailed them at 15 per cent. profit; how much was the gain?

```
                                    7500.00
15 per cent. is written thus,           .15
                                  ---------
                                  1125.0000
```

Ex. 8. Bought land for $25000 and sold it at 25 per cent. advance; how much was made by the operation?

```
         100 per cent. is written thus, 1.00
          25   "    "    "      "        .25
         -----------------------------------
         100   "   and   "      "       1.00
Adding                        25000.00
25 per ct.                        1.25
makes                       ------------
                            31250.0000
                            25000
                            ------------
                            $6250 ans.
```

Ex. 9. Bought land amounting to $110,000, and keeping it 11 years, and sold it at fifty per cent. advance; allowing money to be worth 6 per cent., how much was made?

```
         100 per cent. is written thus, 1.00
          50   "    "    "      "        .50
         -----------------------------------
         150   "   and   "      "       1.50
Adding                      110000.00
50 per ct.                       1.50
makes                       ------------
                            165000.0000
                                     11 years.
                            ------------
                             72600.0000
                            110000
                            -------
                            182600
                            1650000
                            -------
                            $17600 ans.
```

Point off the last 3 examples the same as Ex. 10 in Percentage, page 4.

Ex. 10. A drover bought 46 horses at $85 a piece; 70 cows at $25 a piece; 72 oxen at $28 a piece; what must he sell the whole at, a piece, in order to make $200?

```
                         46×85=3910
                         70×25=1750
                         72×28=2016
                         ----------
                         188    )7676(40.829
                                 752
                                 ----
                                  1560
                                  1504
                                  ----
      188)20000                    560
          -----                    376
          1.06.3                   ---
         40.82.9                   1840
         -------                   1692
        $41.89.2                   ----
```

We now see that the decimal places in the dividen d exceed those in the divisor by 3, counting the ciphers annexed, and none in the divisor; hence we point off 3 decimal places in the quotient; all the figures at the left hand of the point are dollars, and all at the right will be cents and mills.

Ex. 11. Bought 110,000 yards sheeting for 8 cents per yard, and sold the same at 40 cents per yard; what profit was made on the purchase?

```
      110000                 110000
         .08 cts. per yd.       .40 cts. per yd.
      ------                 -------
     8800.00                 4400000
                               8800
                             -------
                             $35200 ans.
```

There are 2 decimals in one factor; we point off 2 decimals in the product; those at the left hand will be dollars.

Ex. 12. A bought 1250 bbls. beef at $10.50 per bbl., and sold it at a loss of 10 per cent.; how much did he lose, and what did he get a bbl.? Ans. 9.45 per bbl. Lost $1312.50.

```
100 per cent. is written thus, 1.00      1250
 10   "     "    "     "      .10       10.50
----------------------------------    --------
 90   "    and   "     "      .90     13125.00
               13125                        .90
               11812.50               ----------
               --------              11812.50.00
                1312.50

                      1250)11812.50
                           --------
                               9.45
```

Deduct 10 per ct. makes

There are two decimals in one of the factors, and 4 in the next two. Point off in the first product the same as Ex. 5, first product, No. 1, page 20, and in the second product the same as Ex. 8, page 3.

We now see that the decimal places in the dividend exceed those in the divisor by 2; hence we point off from the quotient 2 decimals; the figures at the left hand will be dollars, and those at the right hand will be cents and mills.

Ex. 13. Bought candles at 16 cts. 7 mills per lb., and sold them at 20 cts. per lb.; what profit will be made by laying out $100?

```
20   sold for        3.3
16.7 cost.           100
----                ----
 3.3           16.7)330.0
                    --------
                    $19.760 ans.
```

The decimal places in the dividend exceed those in the divisor by 3, counting the ciphers annexed; hence we point off 3 decimals from the quotient; the the figures at the left of the point will be dollars; those at the right will be cents and mills.

Ex. 13. Bought goods for $5025 and sold them on 6 months credit at 22 1-2 per cent. above cost; what was the profit, allowing 7 per cent. interest?

Ans. $954.75.

```
100 per cent. is written thus, 1.00
 22 1-2  "       "      "    .22 5
-----------------------------------
122 1-2  "  and  "      "   1.22.5
```

Adding 22 1-2 per cent. makes

No. 1.	No. 2.	
5025.00	5025.00	
1.225	30	
6155.62500	6)150.75000	int. at 6 per ct.
5200.875	25.12500	
$954.750	175.875	" 7 "
	50.25	
	$5200.875	

Point off the product of No. 1 the same as Ex. 2, page 11, and in No. 2 according to the rule in Simple Interest.

To ascertain at what price merchandise must be sold to gain or lose a stipulated price.

Rule.—1st, Multiply the cost by the rate per cent., and in the product point off two decimal places. The result will be the whole gain or loss. 2d. If a gain, add it to the cost, and if a loss, deduct it therefrom, and you will obtain the selling price.

Ex. 14. A merchant bought cloth for 50 cents per

yard, wishes to mark it so as to gain 12 per cent.; what price must he put on them?

50 cts. purchase price.	50
.12 per cent. profit.	6
6.00	56 cts. selling price.

Must sell them for the purchase price, together with per cent. of that price.

Ex. 15. Bought broadcloth at $4.50 per yard; how much must it be sold at, per yard, so as to lose 15 per cent.?

450 purchase price.	450
67.5	.15 per ct. loss.
$3.82.5 ans.	67.50 amt. loss.

Ex. 16. If I buy 1650 yards flannel for $379.50, how must I retail it, per yard, to gain 25 per cent.?

23 cents per yard.
.25 per cent. profit.

5.75

23 cts cost.	1650)379.50(0.23
5.7	3300
Ans. 28.7	49.50
	49.50

The decimal places in the dividend exceed those in the divisor by 2, hence we point off 2 decimals in the quotient for cents.

To make 10 per cent, on goods that cost $5.00, the merchant should ask $5.50 for them, no matter whether the price of the goods falls or raises on his hands, or remains the same as it was when he bought them.

TO FIND THE FIRST COST.

RULE.—Divide the selling price by the increased rate per cent. and the quotient will be the purchase price.

Ex. 1. A merchant sold broadcloth at $5.40 per yard, and gained 20 per cent.; how much did the cloth cost? Ans. $4.50.

100 per cent is written thus, 1.00
20 " " " " .20

Adding 20 per ct. makes 120 " and " " 1.20

1.20)5.40(4.50
480
600
600

The decimal places in the dividend exceed those in the divisor by 2, counting the ciphers annexed; we point off 2 decimal places in the quotient; the figures at the left hand will be dollars, and those at the right will be cents.

Ex. 2. Sold cloth for $2.10 per yard, by which was lost 30 per cent. on the prime* cost; what was the prime cost? Ans. 3.10.

100 per cent. is written thus, 1.00
30 " " " " .30

Deducting 30 per ct. makes 70 " and " " .70

70)2.17
3.10

Point off the quotient the same as Ex. 1.

*Prime cost means the first cost.

Ex. 3. In order to sell coffee at $6.50 per hundred pounds and make 10 per cent., what must be my purchase price?

Add 10 per ct. makes

100 per cent, is written thus,					1.00
10	"	"	"	"	.10
110	"	and	"	"	1.10

1.10)6.50

5.90.9 ans.

The decimal places in the dividend exceed those in the divisor by 3, counting the ciphers annexed; hence we point off 3 decimal places in the quotient; those at the left will be dollars and at the right will be cents and mills.

PARTNERSHIP.

Partnership is the association of two or more persons in business. The money employed is called the capital or stock, and the profit or loss to be shared among the partners is called the dividend.

To find the gain or loss when the stock of each is employed for the same time—

RULE.—As the whole stock is to each partner's stock, so is the whole gain or loss to each partner's stock.

Ex. 1. A, B and C enter into trade; A put in $500, B put in $700, and C put in $800. They gained $400; what is each ones part of the gain?

500 A put in
700 B "
800 C "

In trade, $2000 amount of stock.

As 2000 : 500 :: 400
400
2000)200,000
100

As 2000 : 800 :: 400
400
2000)320,000
160

As 2000 : 700 :: 400
400
2,000)280,000
140

Dolls.
100 A's gain.
140 B's "
160 C's "
$400

Ex. 2. Five persons, A, B, C, D and E, are to share between them $2400; A is to have 1-6, B is to have 1-4, C is to have 3-8, D and E are to divide the remainder in proportion to the numbers 5 & 7; how much does each one receive?

A receives 1-6 of 2400=$400
B " 1-4 " = 600
C " 3-8 " = 900
1900

5 represents D's part.
7 " B's "
12 " the sum.

Hence D receives 5-12 of 500=208.33 1-3
E " 7-12 " =291.66 2-3
$500.00

Ex. 3. A and B venture equal stock in trade and clear $164; by agreement A was to have 5 per cent. of the profits, because he managed the business. B was to have 2 per cent.; how much was each one's gain, and how much did A receive for his trouble?

Ans. A's gain $117,143
B's " 46,857

```
 164              164
 .05              .02 per cent.
----             ----
 820              328
 328
----
1148
```

Operation by Proportion.

```
1148 : 820 :: 164                        1148 : 328 :: 164
        164                                      164
      -----                                     -----
1148)134480(117,143 nearly.             1148)53792(46,857
     1148                                     4592
     ----                                     ----
      1968                                     7872
      1148                                     6888
      ----                                     ----
       8200     117,143                         9840
       8036      46,857                         9184
       ----     -------                         ----
        1640    $70,286 A received)              6560
        1148      (for his trouble.              5740
        ----                                     ----
         4920                                     8200
         4592                                     8036
         ----                                     ----
          3280
```

Point off the same as Ex. 10, in Loss and Gain, page 22.

Ex. 4. A and B. bought and sold wool; A purchased to the amount of $7840, and paid $250 expenses; B purchased to the amount of $4900, and paid

$288 expenses; A disposed of the wool for $12,475; how must A and B settle the loss to be shared equally?

7840 A purchased. 250 " expenses.		4900 B purchased. 288 " expenses.
8090 5188	13278 12475	5188 313.749
13278	$803 loss.	$4874.251 amt. A pays B.

Operation by Proportion.

13278 : 5188 :: 803.
803

13278)4165964

313.749

B's loss substracted from his stock will be what A is to pay B.

Ex. 5. A and B enter into partnership for one year; A furnishing $1500, B $1000, and receiving $150 for overseeing; how much did B receive? They gained $2256; what was each one's share of the gain?

2256
Deduct 150

2106 bal. of gain to be divided.

Operation by Proportion.

2500 : 1000 :: 2106 1000	1500 1000
2500)2106000	2500 amt. whole stock.
842.40	

1263.60 A's share. 842.40 B's share.		2500 : 1500 :: 2106 1500
$2106.00	842.40 150	2500)3159000
	$992.40 amt. B receives.	1263.60

Point off in the quotients, Nos. 4 and 5, as Ex. 3.

Ex. 6. Divide $360 into 4 equal parts which shall be to each other as 3, 4, 5 and 6.

```
3=3 18=1-6 of 360= 60 ⎫
4=4-18=2-9  "    = 80 ⎬ Ans.
5=5-18=5-18 "    =100 ⎪
6=6-18=1-3  "    =120 ⎭
18                $360
```

To find the gain or loss when their capital is employed for different times—

RULE.—Multiply each partner's stock by the time it is in trade; then say as the sum of all the products is to each particular product, so is the whole gain or loss to each man's share of the gain or loss.

Ex. 7. A and B enter into partnership; A furnishing $500 for 4 months, and B $700 for 5 months; they gained $275; what is each one's share of the gain?

```
 500
   4 months in trade.
 ----
2000                      700
3500                        5 months in trade.
----                      ----
5500                      3500
As 5500 : 2000 :: 275         As 5500 : 3500 :: 275
              2000                              3500
            ------    $100 A's gain.          ------
    5500)5500 00      $175 B's  "      5500)962500
         -------      ----                  ------
             100      $275                     175
```

Ex. 8. A B. and C. enter into partnership; A put in $85 for 8 months, B put in $60 for 10 months, and C put in $120 for 3 months, by misfortune they lost $51; what must each man sustain of the loss;

```
680        85          60          120
600         8 mos.     10 mos.       3 mos.
360        ---         ---          ---
----       680         600          360
1640 sum of products.
```

1640 : 680 :: 41	1640 : 600 :: 41	1640 : 360 :: 41
41	41	41
1640)27880	1640)24600	1640)14760
17	15	9

$17 A's loss.
$15 B's "
$ 9 C's " } Ans.
41

Ex. 9. A and B trade in company for one year only; on the first of January A put in $1200, but B could not put in any money into the stock until the 1st of April; what did he then put in to have an equal share with A?

B 9 mos. in trade : A 12 mos. in trade :: 1200
12
9)14400
$1600 Ans.

The answer to the question is given in money; place the dollars for the 3d term. The answer is to be more, place the next greater number (12) for the 2d term, and the less number (9) for the 1st term. See rule in Simple Proportion.

Ex. 10. Two men engaged in partnership for four years; A put into the firm $12.500; B put in $2500. B is to superintend the business and is paid $10,000, the difference between his and A's capital. At the end of the year A increased his capital to $25,000. They gained by trade $10,655; what is each one's share of the gain?

```
12500                              2500
 2500                             10000
-----                             -----
10000 difference between A &      12500
               B's capital.           4 yrs. in trade.
                                  -----
                                  50000

      12500×4 yrs.=50000
      12500×3  "  =37500
     ---------------------
     $25000        87500 A's product.
                   50000 B's     "
                  ------
                  137500

137500 : 87500 :: 10655
                  87500  137500 : 50000 :: 10655
          --------------                    50000
   137500)932312500                        -------
          ---------------       137500)532750000
              6780.455 nearly.          -----------
              3874.545                    3874.545
              --------
              10655.000
```

Point off in the quotients the same as Ex. 10, p. 22.

Ex. 11. Three men engaged in partnership for 18 months; A put into the firm $3200, and at the end of 4 months he put in $500 more, but at the end of 14 months he took out $800. B at first put in $2200, but at the end of 9 months he took out $1100, and at the end of 12 months he put in $2000. C put in $1500; at the end of 6 months he put in $1500 more, and at the end of 12 months he put in $1500 more, but at the end of 14 months he took out $1000. They gained $3260; what is each man's share of the gain?

Ans. A's share, 1307.40.7
B's " 887.92.9
C's " 1064.66.4

A put in
3200
18 mos. in trade.
———
57600
7000 A ad'd at the end 4 mos.
———
64600 (14 mos.
3200 A took out at end of)
———
61400 A's product.
41700 B's "
50000 C's "
———
153100 sum of products.

B put in
2000
6 mos. in trade.
———
12000

C put in
1500
18 mos. in trade.
———
27000
9000
18000
———
54000
4000 C took out.
———
50000

A put in
500
14 mos. in trade.
———
7000

A took out.
800 (trade.
4 mos. out of)
———
3200

B put in
2200
18 mos. in trade.
———
39600
9900
———
29700
12000
———
41700
1100 B took out.
9 mos out of trade
———
9900

C put in
1500
12 mos. in trade.
———
18000

C put in
1500
6 mos. in trade.
———
9000

C took out
1000
4 mos. out of trade.
———
4000

Operation by Proportion.

153100 : 50000 :: 3260
50000

153100)163000000

1064.66.4 nearly.

153100 : 61400 :: 3260
3260

153100)200164000

1307.407 nearly.

153100 : 41700 :: 3260
3260

153100)135942000

887,929

Point off in the quotients the same as Ex. 10, p. 31.

1307407 A's gain.
1064664 B's "
887929 C's "

$3260,000

COMMISSION AND BROKERAGE.

Commission and Brokerage is the per cent. or sum charged by agents for their services in buying and selling goods, or transacting other business. To compute commission brokerage, discount or stocks—

RULE.—Multiply the given sum by the decimal which expresses the rate per cent.

Ex. 1. A sold goods amounting to $1432.36; how much is the commission at 4 per cent.? Ans. $57.29.

1432.36
.04

$57.29.44

Point off in the same as Ex. 8, in Percentage, p. 3.

Ex. 2. What is the brokerage on $6200 at 3-4 per cent.? Ans. $46.50.

1st method.	2d method.
6200.00	6200.00
.0075=3-4	.01=1 per ct.
46.50.0000	4-2)62.0000 at 1 per ct.
	31.0000=1-2
	15.5000=3-4
	$46.50.00 ans.

In the first method there are 6 decimals in both factors, and in the 2d method 4 decimals in both factors. In the 1st method we point off from the product 6 decimal places; the figures at the left will be dollars, those at the right will be cents. In the 2d method the same as Ex. 2, p. 9. The decimal places in the dividend exceed those in the divisor by 4; hence we point off 4 decimal places in the quotients; those at the left hand will be dollars, and those at the right will be cents.

Ex. 3. Engaged a broker to purchase 12 shares of Boston & Maine Railroad at $112.50 per share; what is the commission at 1-4 per cent.?

1st method.	2d method.
112.50	112.50
12 shares.	12 shares.
1350.00	1350.00
.0025=1-4	.01
3.37.5000	4)13.50.00 at 1 per ct.
	3.37.50

In the 1st method there are 2 decimals in one of the factors and 6 in the next two. We point off in the 1st product 2 decimals; all at the left hand will be dollars. In the 2d product 6 decimals; those at the left will be dollars, and those at the right will be cents and mills. In the 2d method we point off in the 1st product the same as we did in the 1st product in the 1st method, and in the 2d product the same as Ex. 8, p. 3. The decimal places in the dividend exceed those in the divisor by 4; hence we point off four decimal places in the quotient—those at the left will be dollars, and those at the right will be cents and mills.

Ex. 4. What must be paid to a New York broker for $5000 of City Bank bills on Eastern Banks at 1-4 of 1 per cent.?

```
   5000.00              5000
       .01                12.50
   -------              -------
4)50.00.00=1 per ct.    $5012.50 ans.
  -------
   12.50
```

Ex. 5. A drover exchanges $2240 of country money for city bills, paying 1-8 per cent. on his country money; what does he receive?

```
   2240.00
       .01              2240
   -------                 2.80
8)22.40.00 at 1 per cent.  -------
   -------              $2237.20 ans.
     2.80
```

Ex. 6. A sells for an individual 90 shares of Boston & Worcester Railroad stock for $125 a share, receiving

1 per cent. on what money he gets; what does he receive?

```
  125.00 a share.
      90 shares.
  -------
11250.00
     .01=1 per cent.
---------
$112.50.00
```

Point off in the 1st product the same as in 1st method Ex. 3, 1st product, and in 2d product 4 decimals; the figures at the left will be dollars, and those at the right will be cents.

Ex. 7. A sold 12 pieces of cloth for B containing 28 yards, at $3.75 per yard, and charged 2 1-2 per cent. commission and 3 per cent. guaranteeing the payment; how much will A receive?

```
No. 1.
    28
    12                     No. 2.
  ----                     1260.00
   336                        .055=5 1-2
  3.75                    ---------
 -------                  69.30.000
1260.00
  69.30
 -------
1190.70 Ans.
```

In No. 1 we point off in the product the same as Ex. 3, 1st method, 1st product; in No. 2 the same as Ex. 6, Insurance, p. 12.

Ex. 8. A merchant in Chicago consigned 450 bbls. of beef at $11.25 per bbl., and charged 2 1-2 per cent. commission; the merchant sold a draft on his agent for the sum due him at 1 per cent. premium; how much did he receive for his beef?

```
        100 per cent. is written thus, 1.00
          1   "    "    "     "     .01
        ---------------------------------
        101   "   and   "     "     1.01
Adding
1 per ct.      No. 1.                  No. 2.
makes          11.25                  5032.50
                 450                   126.56
               -------                -------
               5062.50                4935.94
                  .025=2 1-2             1.01
             -----------              ----------
             $126.56.250             4985.29.94 Ans.
```

In No. 1 we point off in the 1st product the same as Ex. 3, p. 36, 1st method, 1st product, and in the 2d product the same as Ex. 6, p. 16, and in No. 2 the same as Ex. 8, p 3.

Ex. 9. A manufacturer in Albany sent 256 pieces of cloth, containing 28 yards in a piece, to Wm. F. Smith of Boston, and agreed to pay him 2 per cent. commission and 3 per cent. for guaranteeing. S. sold the cloth at $4.65 per yard, paid $28.35 freight and $15.17 insurance; how much did A receive for his cloth? Ans. $31621.12.

```
      256             33331.20            1666.56
       28              1710.08              28 35
     -----            ---------             15.17
     7168 yds         31621.12 ans.       --------
     4.65 per yd.                         $1710.08
   --------
   33331.20                                      (teeing.
      .05=5 per cent. for commission and guaran-)
   ----------
   1666.56.00
```

Point off in the 2d and 3d products the same as Ex. 3, 2d method, p. 36.

COMMISSION DEDUCTED IN ADVANCE.

If the commission is to be deducted from the given sum, it is evident that we ought not to pay an agent commission on his own money, which, however, is often unjustly practiced.

RULE.—Divide the given sum by the increased rate per cent., the quotient will be the amount to be invested; substract the amount from the given sum and you have the required commission.

Ex. 1. A sent his agent $1200 to purchase wool, how much had he ought to pay out, after deducting his commission of 5 per cent.; what was his commission?

100 per cent. is written thus, 1.00
5 " " " " .05
105 " " " " 1.05

Adding 5 per ct. makes

1st method
1.05)1200.00(1142,857
105
150
105
450
420
300
210
900
840
600
525
750
735

2d method.
105 : 5 :: 1200
5
105)6000(57,142
525
750
735
150
105
450
420
300
210

1200
1142.85.7
$57,143 Ans.

See rule in Division of Decimal Fractions.

Where the decimal places in the dividend exceed those in the divisor, make them equal by annexing ciphers to the right of the dividend, and the quotient will be a whole number.

We now see that the decimal places in the dividend exceed those in the divisor by 3, counting the ciphers annexed; hence we point off 3 decimal places in the quotient. The figures at the left of the point will be dollars, and those at the right will be cents and mills.

Ex. 2. H. Kinsman sold for Benj. Billings of Rutland, 3600 lbs. butter at 20 cents per pound; 2670 lbs. cheese at 8 cents per lb., at a commission of 5 per cent. He invested the balance in books, after deducting his commission of 2 1-2 per cent., for purchasing; what amount of goods ought he to receive?

```
100 per cent. is written thus, 1.00
  2 1-2  "     "    "     "    2.5
-----------------------------------
102 1-2  "    and   "     "  1.02.5

Add 2 1-2 per ct. makes

                              1.02.5)886,920
                                     --------
      3600                           $865.287 ans.
       .20
    ------                  933.60
    720.00                   46.68
    213.60                 -------
    ------                 $886.92
    933.60
       .05                            2670
    ------                               8
  46.68.00 commission.               -----
                                    213.60
```

We annex one cipher to the right of the dividend and point off in the quotient in the same manner as Ex. 1, p. 41.

Ex. 3. Wm. F. Grooms sent a broker $15000 to invest in goods; after deducting his commission of 1 3-4 per cent.; what amount of goods ought he to receive?

100 per cent. is written thus, 1.00
1 3-4 " " " " .0175

101 3-4 " and " " 1.01.75

Add 1 3-4 per ct. makes

1.0175)15000.0000

14742.014 Ans.

Ex. 4. A man sent a broker $10478.13 to lay out in stocks after deducting his brokerage at 1-2 per cent.; what was the brokerage, and how much stock did he receive?

100 per cent. is written thus 1.00
1-2 " " " " 0.5

100 1-2 " " " " 1.00.5

Add 1-2 per ct. makes

1.005)10478.130

$10426 his stock.

10478.13
10426.

$52.13 brokerage.

The last two examples are worked in the same manner as Ex. 1.

SIMPLE INTEREST.

There are three things to be mentioned in Interest;

1st. The principal, or money lent.

2d. The rate or sum per cent. agreed on.

3d. The amount, or principal and interest added together.

Interest is of two kinds, Simple and Compound. Simple Interest is that which is allowed for the principal. Compound Interest is that which arises from the interest being added to the principal and continuing in the hands of the lender and becomes a part of the principal at the end of each stated time of payment.

RULE.—Write down half the greatest even number of months for a multiplier—there being an odd month it must be reckoned 30 days—add the given days, if any; seek how many times you can have 6 in the sum of them; place the figure for a decimal at the right hand of half the even number of months already found, by which multiply the principal, observing in pointing off the product to remove the decimal point two figures from its natural place towards the left hand; that is, point off two more places for decimals in the product than there are decimal places in the multiplicand and multiplier counted together. Then all the figures at the left of the point will be dollars, and those at the right will be cents and mills.

Should there be a remainder in taking 1-6 of the days, reduce it to a regular fraction, for which take aliquot parts of the multiplicand. Thus—

If the	remainder be	1=	divide	the	multiplicand	by	6
"	"	2=⅓	"	"	"	"	3
"	"	3=½	"	"	"	"	2
"	"	4=⅔	"	"	"	"	3 twice
"	"	5=½&⅓	"	"	"	"	2 & 3

The quotients which in this way occur, must be added to the product of the principal, multiplied by half the months, &c. When there are days less number than 6, so that 6 cannot be contained in them, put a cipher in place of the decimals at the right hand of the months, then proceed in all as above directed.

To compute interest when the rate is greater or less than 6 per cent.

RULE.—First find the interest on the given sum at 6 per cent.; then add to this interest or substract from it such fractional part of itself as the given rate exceeds or falls short of 6 per cent.

When the required rate is 6 1-2 per cent., we first find interest at 6 per cent. and add 1-12 of it to itself.

When the required rate is 7 per cent., we first find the interest at 6 per cent. and add 1-6 of it to itself. If 7 1-2 per cent., we add 1-4 of it to itself.

When the required rate is 8 per cent., we first find the interest at 6 per cent. and add 1-3 of it to itself.

When the required rate is 8 1-2 per cent., we first find the interest at 6 per cent. and add 5-12 of it to itself.

When the required rate is 9 per cent., first find the interest at 6 per cent. and add 1-2 of it to itself.

If 9 1-2 per cent., find the interest at 6 per cent. and add 7-12 of it to itself.

If 10 per cent., find the interest at 6 per cent. and add 2-3 of it to itself.

If 10 1-2 per cent., find the interest at 6 per cent. and add 3-4 of it to itself.

If 11 per cent., find the interest at 6 per cent. and add 5 6 of it to itself.

If 11 1-2 per cent., find the interest at 6 per cent. and add 11-12 of it to itself.

If 12 per cent., find the interest at 6 per cent. and multiply it by 2.

If 1 per cent., we find the interest at 6 per cent. and substract 5-6 of it from itself.

If 1 1-2 per cent., we find the interest at 6 per cent. and substract 3-4 of it from itself.

If 2 per cent., we find the interest at 6 per cent. and substract 2-3 of it from itself.

If 2 1-2 per cent., we find the interest at 6 per cent. and substract 7-12 from itself.

If 3 per cent., we find the interest at 6 per cent. and substract 1-2 from itself.

If 3 1-2 per cent., we find the interest at 6 per cent. and substract 5-12 from itself.

If 4 per cent., we find the interest at 6 per cent. and substract 1-3 from itself.

If 4 1-2 per cent., we find the interest at 6 per cent. and substract 1-4 from itself.

If 5 per cent., we find the interest at 6 per cent. and substract 1-6 from itself.

If 5 1-2 per cent., we find the interest at 6 per cent. and substract 1-12 from itself.

Ex. 1. What is the interest of $76.54 for 1 year, 7 months and 11 days at 6 per cent.? Ans. $7.41.1.

```
3.2)76.54
      9.6
    -----
    45924
   68886
     3827=1-2
     2557=1-3
   ---------
  7.41.168
```

Note.—The number of months being 19 the greatest even number is 18, 1-2 of which is 9, which write down, then seeking how often 6 is contained in 41, (the sum of the days in the odd month and given days), we find it will be 6 times, which also set down at the right haud of 1-2 of the even number of months for a decimal, by which, together, we multiply the principal. In taking 1-6 of the days (41) there will be a remainder of 5=1-2 and 1-3, for which we take first 1-2 the multiplicand or principal; that is, divide the multiplicand by 2, then by 3, and these quotients added, with the products of 1-2 the number of months, &c., the sum of these will show the interest required; observing to count off 2 more figures for decimals in the product than there are decimal figures in both the multiplier and multiplicand counted together.

Ex. 2. What is the interest of $375.58 for 6 months at 6 per cent.? Ans. $11.26.7.

```
   375.58
      3.0
 ---------
$11.26.7.40
```

Put a cipher in the place of the decimals at the right hand of half the even number of months. See rule, page 43.

Ex. 3. What is the interest due on a note of $523 from March 5th, 1869, to Aug. 20, 1870, at 7 per cent.?

Ans. $53.38.9

yrs.	mos.	dys.	
1870	8	20	2)523.00
1869	3	5	8.7
—	–	—	——
1	5	15	366100
12		30	418400
2)17 mos.		6)45	26150
—		—	——
			6)45.76.250 int. at 6 per cent.
8		7	7.62.708
			——
			$53.38.958 int. at 7 per cent.

See rule for pointing off in the products, p. 44.

Ex. 4. What is the interest of $800 for 4 months and 3 days at 5 per cent.? Ans. $21.86.

```
2)800.00
     2.0
 --------
 1600000
   40000
 --------
3)16.40000 int. at 6 per cent.
   5.46666
 ----------
$21.83.666
```

Ex. 5. What is the interest of $240.38 for 7 months and 10 days at 5 per cent.? Ans. $7.34.4

```
        30        3.3)240.38
2)7     10               3.6
 -      --           -------
 3-   6)40           144228
        --           72114
        6-4            8012
                       8012
                     -------
                   6)8.81392 int. at 6 per cent.
                     1.46898
                     -------
                     7.34.494 int. at 5 per cent.
```

Ex. 6. What is the interest of $462 for 3 years at 6 per cent.? Ans. $83.16.

```
462 00
   .06
-------
27.72 00 int. for 1 year.
       3 years.
--------
83.16.00 int. for 3 years.
```

Ex. 7. What is the interest of $550 for 1 year at 7 3-10 per cent.?

```
550.00
   7.3=7 3-10
-------
 165000
385000
-------
$40.15.000 Ans.
```

Ex. 8. What is the interest of $746.28 for 1 year, 4 months at 10 per cent.?

```
746.28
     8 0=half the mos.    2 of 5970=11940
-------                   -----------------
                          3            3
59.70.240
39.80                          3)11940
-------                         ------
$99.50 int at 10 per cent.      39.80
```

Ex. 9. What is the interest of £587 16s. 3d. for 3 years at 6 per cent.? Ans. £105 16s. 1d.

RULE.—Reduce the shillings and pence to a decimal of a pound, then compute the interest as though the sum were dollars and cents, and finally reduce the decimal figures in the answer to shillings and pence.

```
16.3        240)19500      587.8125
12          -----              .06
---          8125*         ---------
195                        35.268750 for 1 year.
---                                3
240                        ---------
                          105.806,250 for 3 years.
               .806 decimal of a pound.
                 20 shillings a pound.
             ------
             16,120
                 12 pence a shilling.
              -----
              1,440
```

Ex. 10. What is the interest of $700 for 5 years at 1 per cent. per month? Ans. $35.

```
700.00
   .01=1 per cent.
-------
7.00.00 interest 1 month
      5 months.
-------
35.00.00
```

*Reduced to a decimal of a pound.

Ex. 11. What is the interest of $1000 for 1 year, 9 months at 2 per cent. a month? Ans. $420

```
 1000.00
      .02
 -------
 20.00.00
       21 months.
 -------
$420.0000
```

Ex. 12. What is the interest of $100,000 for 3 years at 3 per cent. a month? Ans. $108,000.

```
 100000.00
      1.08
 ---------
  80000000
 10000000
 ---------
108000.0000
```

Three per cent. a month is 36 per cent. a year, and for 3 years is 108 per cent. Point off in the products in the same manner as Ex. 11, in Percentage, p. 4.

RULE.—For casting interest by multiplying the principal by the number of days. Reckon 30 days to a month, 360 days a year. If there are no cents in the principal, annex two ciphers in place of them. Divide the product by the following numbers, because it takes so many days for $1 to gain $1.

If 3 per cent. divide the product by 12000
" 4 " " " " " 9000
" 5 " " " " " 7200
" 6 " " " " " 6000
" 7 " " " " " 5143
" 8 " " " " " 4500
" 9 " " " " " 4000
" 10 " " " " " 3600

Point off in the quotient two figures; the figures at the left hand will be dollars--those at the right hand will be cents and mills.

Ex. 1. What is the interest of $268 for one year, 2 months and 7 days at 6 per cent.?

```
360 days a year.                26800
 60 days in 2 months.             427
  7 days.                  ----------
---                  6,000)11443,600
427                        ----------
                           $19.07 ans.
```

Ex. 2. What is the interest of $58.28 for 1 month and 3 days at 5 per cent.?

```
          5828
            33
        ------
  7200)192324
        ------
            26 cents, ans.
```

Ex. 3. What is the interest of $800 for 12 days at 7 per cent.?

```
          80000
             12
        -------
  5143)960000
        -------
           1.86
```

DISCOUNT BY COMPOUND INTEREST,

Is the present worth of a debt payable at some future time without interest, is the sum which, being put at legal interest, will amount to the debt at the time it becomes due.

RULE.—Divide the debt by the amount of $1, for the given time, and the present worth, which, if substracted from the given sum will leave the discount.

Ex. 1. What is the present worth of $600, due 3 years hence, at 6 per cent., compound interest; what is the discount?

The amount of $1 for 3 years at Compound interest is 1,191016)600,000000 600,000 given sum.

503,771 present worth.

See table, p. 53. $503.771

96.229 discount.

Ex. 2. What is the present worth of $500, due 4 years hence, at 7 per cent., Compound Interest?

Ans. $381,447.

The amount of $1 for 4 years, Compound Interest, at 7 per cent. is 1,310796)500,000000

$381.447

Annex ciphers to the right of the dividend and point off in the quotient in same manner as Ex. 1., Commission Deducted in Advance, p. 40.

TO FIND THE PRINCIPAL FROM THE INTEREST AND THE RATE PER CENT.

Rule.—Divide the given interest by the interest of $1 for the given time and rate per cent.

Ex. 1. What principal will, in 1 year, 7 months and 15 days, at 6 per cent., give $9.75 interest?

```
2)1.00                        9.75
   9.7                         100
  ----                        ----
  9700                   975)97500
    50                        ----
  ----                     $100 ans.
.09.750=9 cts. 7 1-2 mills int.
```

Ex. 2. Mexico pays England, France and Spain $3,000,000 interest a year; what principal will, in one year, at 6 per cent., give that interest.

```
  1.00                                   3000000
   .06                                       100
 -----                                  --------
.06.00=6 cents interest.              6)300000000
                                       ---------
                                     $50,000,000
```

Ans., Fifty millions of dollars.

TO FIND THE TIME FROM THE PRINCIPAL AND THE RATE PER CENT.

RULE.—Divide the given interest by the interest of the given principal for 1 year at the given rate per cent.

Ex. 1. In what time will $700, at 7 per cent., give $85.75? Ans., 1 year, 9 mos.

```
    700.00                 8575
       .07                  100
  --------               -------
  49.00.00               )857500(1
                          490000
                         -------
                          367500
                              12 mos. a year.
                         --------
                         4410000(9
                         4410000
```

Ex. 2. In what time will $50, at 6 per cent., give $2.05? Ans., 8 mos., 6 days.

```
    50.00              205
      .06              100
  -------             -----
  3.00.00        30000)20500
                          12 mos. a year.
                      ------
                      246000(8
                      240000
                      ------
                        6000
                          30 days a month.
                      ------
                      180000(6
                      180000
```

COMPOUND INTEREST TABLE,

Showing the amount of $1 at 5, 6, 7 and 8 per cent. from 1 to 12 years.

Yrs	5 per cent.	6 per cent.	7 per cent.	8 per cent.
1	1,050,000	1,060,000	1,070,000	1,080,000
2	1,102,500	1,123,600	1,144,900	1,166,400
3	1,157,625	1,191,016	1,225,043	1,259,712
4	1,215,506	1,262,476	1,310,796	1,360,489
5	1,276,281	1,338,225	1,402,552	1,469,328
6	1,340,095	1,418,519	1,500,730	1,586,874
7	1,407,100	1,503,630	1,605,781	1,713,824
8	1,477,455	1,593,848	1,718,186	1,850,930
9	1,551,328	1,689,479	1,838,459	1,999,005
10	1,628,894	1,790,847	1,967,151	2,158,925
11	1,710,339	1,898,298	2,104,852	2,331,639
12	1,795,856	2,012,196	2,252,191	2,518,170

Ex. 1. What is the compound interest of $425 for 9 years, 2 months and 8 days at 6 per cent.?

Am't of $1 by the table is 1,689,479
425 principal.

3) 718,028575 principal and int.
1.1

8.13,756
718.02

726.15
425

$301.15 ans.

COMPOUND SEMI-ANNUAL INTEREST TABLE,

Sowing the amount of $1 at 5, 6, 7 and 8 per cent., from 1 to 12 years.

Yrs	5 per cent.	6 per cent.	7 per cent.	8 per cent.
1	1,050,625	1,060,900	1,071,225	1,081,600
2	1,103,812	1,125,508	1,147,522	1,169,858
3	1,159,692	1,194,051	1,229,254	1,265,318
4	1,218,401	1,266,768	1,316,807	1,368,567
5	1,280,082	1,343,914	1,410,596	1,480,241
6	1,344,886	1,425,759	1,511,065	1,601,028
7	1,412,970	1,512,587	1,618,690	1,731,671
8	1,484,501	1,604,703	1,733,981	1,872,974
9	1,559,653	1,702,429	1,857,483	2,025 807
10	1,638,610	1,806,106	1,989,781	2,191,112
11	1,721,564	1,916,098	2,131,503	2,369,906
12	1,808,718	2,032,187	2,283,319	2,563,290

Ex. 1. What is the compound semi-annual interest of $300 for 3 years and 10 days at 7 per cent.?

The am't of $1 for three
years by the table is 1.229254
300

3,3)368,77.6200
0.1=1-6 of the days.

33,8.77=36 cents, 8 mills.
12,2.92
12,2.92

6)61,4.61 int. at 6 per cent.
10.243

71,7.04 int. at 7 per cent.
36877

369.48
300 deduct the principal.

$69.48 ans.

PARTIAL PAYMENTS.

RULE.—If the payment exceeds the interest, the surplus goes toward discharging the principal, and the subsequent interest is to be computed on the balance of the principal. If the payment be less than the interest, it must not be taken out of the principal, but interest continued on the former principal until the period when the payments taken together exceed the interest due them, surplus is applied toward discharging the principal.

Ex. 1. What is due on a note of $750, dated Jan. 1st, 1868, with interest at 6 per cent. after deducting the following endorsements:

August 10 1868, paid $110.
April 13 1869, " 115.

What was due on taking up the note Nov. 20, 1869.

Years.	Mos.	Days.	
1868	8	10	
1868	1	1	
	7	9	time to 1st paym't, Aug. 10, '68.

Years.	Mos.	Days.	
1869	4	13	
1868	8	10	
	8	3	time on bal. after 1st paym't, Aug. 10, '68.

Years.	Mos.	Days.	
1869	11	20	
1869	4	13	
	7	7	time on bal. after deducting the last paym't Apr. 13, '69

Principal, - - - - - -	750.00
Int. to 1st paym't, Aug. 10 '68, - -	27,375
	777,375
st paym't to be deducted from amount, -	110
Bal. due after 1st paym't, - - -	667,375

Int. on bal. to Apr. 13, '69, - - -		27,028
		694,403
2d paym't to be deducted from amount,		115
		579,403
Int. due on taking up the note, - -		20,955
Answer,		$600,358

Ex. 2. What is due on a note of $875, dated March 8, 1867, with interest at 7 per cent, after deducting the following endorsements:

Sept. 8, 1867, paid $75.
June 18, 1868, " 15.
Mar. 24, 1869, " 90.

What was due on taking up the note Jan. 9, 1870.

Principal, - - - - -		$875.00
Interest to 1st paym't, Sept. 8, 6 mos.,		30.625
		905.625
1st paym't deducted from amount, -		75
Bal. due after 1st paym't, Sept. 8, '67,		$830.62.5
Int. on bal. 2d paym't, June 18, '68, 9 mos., 10 days, - -	45.222	
2d paym't being less than int. due,	15	
Surplus int. unpaid June 18, '68,	30.222	
Int continued on bal. from June 18, to Mar. 24, '69, 9 mos., 6 days,	44.576	74.798
		905.423
3d paym't being greater than the int. due is to be deducted from the am't. - -		90
		815.423
Int. on bal. to Jan. 9, '70, 9 mos., 15 days,		45.187
		$860.61.0

BANKING.

A bank is a corporation chartered by law for the purpose of receiving money and furnishing a paper currency. It is customary for a bank in discounting a note or draft, to deduct in advance the legal interest on the given sum from the time it is discounted to the time when it becomes due.

Bank Discount is the same as Simple Interest paid in advance; thus, the bank discount on a note of $106, payable in one year at 6 per cent., is $6.36, while the true discount is but $6.

A Teller is a clerk in a bank who receives and pays the money on checks and drafts.

A check is an order for money.

A draft is an order from one man to another, directing the payment of money—a bill of exchange.

Rule.—Compute the interest on the face of the note for 3 days more than the specified time, (these are called days of grace), then calculate the interest at the given rate; the result is the bank discount and the face of the note, diminished by the discount, is the present worth.

Ex. 1. What is the discount, and what is the present worth of a note for $460 at 60 days, discounted at a bank at 6 per cent., and grace.

Ans. $455.17 is the present worth.
and 4.83 " discount.

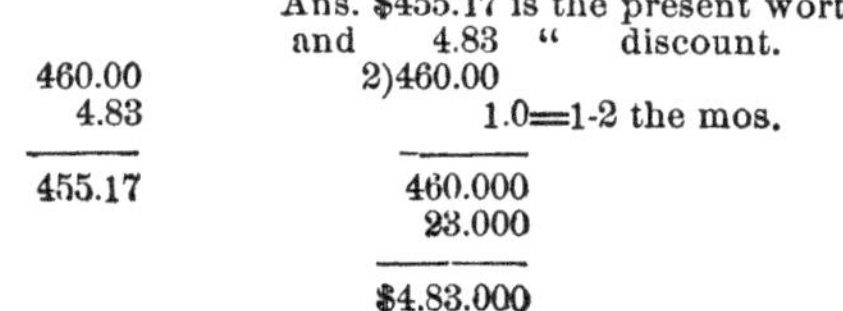

460.00
4.83
455.17

2)460.00
1.0=1-2 the mos.
460.000
23.000
$4.83.000

See rule in Simple Interest for pointing off the decimals in the product, p. 43.

Ex. 2. What is the bank discount on a note at 7 per cent. for $850.64, payable in 90 days and grace?

```
2)850.64                          Ans. 15.38.
     1.5=1-2 the mos.
  ------
  425320
  85064
   42532
  ----------
6)13.18 492  interest at 6 per cent.
  2.19.748
  ----------
  15.38.240  bank discount, or int. at 7 per ct.
```

Ex. 3. What is the present worth of a note at 8 per cent. for $1500, payable in 12 days, and grace discounted, at a bank?

```
                     2)1500.00
6)12 days                 0.2=1-6 of days.
  --                   -------
   2                   30.0000
                        7.5000
1500                   -------
  5.00                3)3.75000 int. at 6 per cent.
--------                1.25
$1495.00 ans.          -----
                        5.00 int. at 8 per cent.
```

Ex. 4. What is the bank discount on a note at 7 3-10 per cent., for $125 for 6 mos. and 10 days?

RULE.—Multiply the principal by 2, that will give the interest for 1 day—because it is 2 cents a day on $100—that product by the number of days.

```
182 1-2 days in 6 months.
 10      "   added
-------
192 1-2 days
```

```
       125.00
          .02
      ───────
    2).02.5000=2 1-2 cents per day.
         192 1-2
      ──────────
      4.81.2500
```

We point off in the 1st product according to the rule in Simple Interest, p. 43. We prefix one cipher in the 1st product because cents occupy 2 places. In the 2d product point off 6 decimals, because there are 6 decimals in the multiplicand. All at the left will be dollars, and those at the right will be cents and mills.

Ex 5. A bought 20400 lbs Rice at 10 cts. per pound, cash, and sold it immediately for 15 cts. per pound on 6 months credit. If he should get this discounted at a bank, what will he gain on the Rice?

```
   1                             2
 20400                         20400
   .10 cts. a pound.             .15
 ───────                       ───────
 20.40 00                      30.6000
    91.80                          3.0=1-2 the mos.
 ───────                       ──────────
  2131 80                      91.80.000
                 3060
                 2131.80
                 ───────
                 $928.20 ans.
```

Point off in two products of Nos. 1 and 2 the same as Ex. 11, in Loss and Gain, p. 17; and the last product of No. 2 according to the rule in Simple Interest, p. 43.

THE PROCEEDS OF A NOTE FIND THE FACE.

It is often required to make a note of which the present value shall be a given number.

RULE.—Divide the proceeds of a note by the proceeds of $1 for the given time and rate mentioned, and the quotient will be the face of the note.

Ex. 1. A merchant wishes to borrow $400 at a bank, for what sum must he draw his note, payable in 60 days, so that when discounted at 6 per cent., he will receive the desired amount?

```
2)100
  1.0                              100
 ----                              .0105
 1000                              -----
   50                              .9895
 -----                                  (from $1.
01.0.50 int. on $1 is 1 cent., 5-10 of a mill; deduct this)
             .9895)400,0000(404,244
                   39580
                   -----
                    42000
                    39580
                    -----
                     24200
                     19790
                     -----
                      44100
                      39580
                      -----
                       45200
                       39580
                       -----
```

Annex ciphers to the right of the dividend and point off in the quotient in the same manner as Ex. 1, in Commission Deducted in Advance, p. 40.

Ex. 2. What is the face of a note at 60 days at 7 per cent. which yields $670, when discounted at a bank?

Ans. $67.318.

```
2)100
  1.0                         100
 ----                         .01225
 1000                         ------
   50                         .98775
 -----
```

6)1.0.5C=1 cent and 5-10 of a mill, int at 6 per cent.
  175

01.2.25=1 cent, 2 mills, 25-100 of a mill int. at 7 per ct.
.98775)670,00000

$678.31 nearly.

Ex. 3. What is the face of a note at 90 days at 5 per cent. of which the proceeds are $1000, when discounted at a bank?

2)100
  1.5=1-2 the mos.

1500
  50

6)1.5.50
  258

100
1292

.98708)1000.00000

Ans. 1013.089

01.2.92 int. at 5 per ct. is 1 ct., 2 mills, 92-100 of a mill

Ex. 4. What is the face of a note at 30 days at 8 per cent., of which the proceeds are $1500, when discounted at a bank?

2)100
  5

500
  50

3)5.50 int. at 6 per ct. is 5 mills and 5-10 of a mill.
  183

7.3.3 int. 7 mills, 33-100 of a mill at 8 per cent.

100
00.733

.99267)1500,00000

Ans. $1511.076

Nos. 2, 3 and 4 examples are worked in the same manner as Ex. 1, p. 60.

EQUATION OF PAYMENTS.

Equation of payments is the equalized or average time when two or more payments, due at different times, may be made at once without loss to either party.

RULE.—Multiply each payment by the time at which it is due, then divide the sum of the products by the sum of all the payments, the quotient will be the answer.

Ex. 1. A owes B. $600 to be paid in 4 mos., $700 to be paid in 6 mos., and 1200 to be paid in 7 mos.; what is the average time for the payment of the whole?

Ans. 6 mos.

Operation.

	$600×4 mos.=2400	
	700×6 " =4200	
	1200×7 " =8400	
Sum of pay'ts,	2500 2500)15000	sum of products.
	Ans. 6 mos.	

Ex. 2. A bought a store for $17380 on 2 years credit; in 8 months he paid $1238; in 4 months more he paid $2217; 3 months later he paid $3369, and 2 months after he paid $1865; how long ought the payment of the balance to be deferred?

RULE.—Multiply each payment by the time it was made before it becomes due, and divide the sum of the products thus obtained by the balance remaining unpaid, the quotient will be the time required. When there are months and days, the months must be reduced to days. Ans. 19 months, 9 days.

```
1238×16=19808
2217×12=26604
3369× 9=30321
1865× 7=13055
-------------
8689    )89788(10
         8689
         ----
          2898
            30 days a month.
         -----
         86940(9
         78201
         -----
```

Ex. 3. A merchant bought the following bill of goods on 6 months credit: Jan. 1, $60; Jan. 19, $80; Jan. 25, 30; Feb. 19, $225, and March 10, $100. At what time must a note for the whole amount be dated so that the buyer shall have 6 mos. credit?

Rule.—Multiply each charge by the number of days from the date of the first charge to its own date; divide the sum of the products by the sum of the charges, and the quotient will be the average time from the first charge.

```
                         days.
          Jan. 1st,  $60×00=00000
            "  19,    80×18= 1440
            "  25,    30×24=  720
          Feb. 19,   225×49=11025
          Mar. 10,   100×68= 6800
-------------------------------------
Sum of products,     495     )19985
                              -----
                              40 days.
```

Therefore the note must be dated Feb. 9, counting from Jan. 1, the $60 will have no time of credit.

Ex. 4. A bought a bill of goods March 11, 1869, to the amount of $1980, on a credit of 4 months, and made the following payments: April 7, $500; May 15, $250, and June 20, 300; when should he pay the balance? Ans. Sept. 22d.

Rule.—Multiply each sum by its time of payment and divide the sum of products by the balance remaining unpaid.

	days.
1980	500×95=47500
1050	250×57=14250
	300×21= 6300
930 bal. unpaid.	1050 930)68050
	73 days.

The equated time for the payment of the above bill is 73 days from July 11th, which will be Sept. 22d, averaging accounts bearing interest at 6 and 7 per cent.

Rule.—Multiply each item of debt and credit by the number of days from its entry to the time of settlement.

Divide the sum of the products on each side by 6000 if 6 per cent., and 5143* if 7 per cent., and the quotients will be the interest due on the respective side.

Finally, substract the amount of the small side from that of the greater and the remainder will be the true balance.

Henry A. Warren in account with Geo. W. Warren.

1869	Feb. 1,	To mdse	350 00	1869	Feb. 20,	By cash,	136 00
"	Mar 15,	" "	255 00	"	May 1,	" "	165 00
"	Apr. 24,	" "	660 00	"	" 5,	" "	217 00

What is the amount due Aug. 1st, 1869, at 6 and 7 per cent.?

*See page 49.

To find the amount due at 6 per cent.

days. *Drs.*	days. *Crs.*
350.00×181=6335000	136.00×162=2203200
255.00×139=3544500	165.00× 92=1518000
660.00× 99=6534000	215.00× 88=1892000
27.35 int.	9.35 int.
6,000)16413,500	6,000)5613,200
$27.35	525.35 $9.35
1292.35	
525.35	

$767.00 ans. at 6 per cent.

To find the amount due at 7 per cent.

350.00	136.00
255.00	165.00
660.00	215.00
31.91 int.	10.91 int.
$1296.91	$526.91
526.91	

$770 00 ans. at 7 per cent.

Sum of products, Dr. side.	Sum of products, Cr. side.
5143)16413500	5143)5613200
$31.91	$10.91

Ex. 5. A bought a bill of goods, amounting to $560, April 3d, on 30 days time, and made the following payments; when in equity should he pay the balance?

April 7, $75
" 12, 30
" 17, 65

75×26=1950
30×21= 630
65×16=1040

170 390)3620(9 days.
3510
110

560
170
390 bal. unpaid.

The equated time for the payment of the above bill is 9 days from May 3d, which will be May 12th. The fraction of a day is never regarded in business operations.

AVERAGE.

RULE.—If the sum of the quantities of different values be divided by the number of those quantities, the quotient is called the *average* of the given quantities, or their mean value.

If there are but two quantities, their mean value is the half sum of the values of those quantities.

Ex. 1 At one of the public schools (of Rutland, Vt.) the number of pupils in attendance on Monday was 694; on Tuesday 675; on Wednesday 681; on Thursday 653; and on Friday 622. What was the average attendance for that week? Ans. 665.

```
                       694
                       675
                       681
                       653
                       622
                      ----
  (5 is the          )3325
 number quantities.   ----
                       665
```

Ex. 2. If No. 1 marble cost 80 cents per superficial foot, and No. 2 cost 45 cents per superficial foot; what is the average cost?

```
                         80
                         45
                         --
      2 is the         )125
 number of quantities.   --
                         62 1-2 ans.
```

AVERAGE OF STORAGE.

RULE.—Multiply the number of barrels or boxes, as the case may be, by the number of days they are in store, and divide the sum of the products by 30, or any other sum agreed upon. The quotient will be the number of articles on which storage is charged for the term.

If the fraction is less than one-half, reject it; but if contains more than one-half, regard it as an entire article.

Ex. 1. What will be the cost for the storage of flour at 6 cents per barrel, which was received and delivered as follows?

Received	May 1,	1869,	2000 bbls.
"	" 26,	"	4000 "
Delivered	May 16,	1869,	1000 bbls.
"	June 1,	"	2000 "
"	" 12,	"	2200 "
"	July 2,	"	800 "

			bbls.	dys.	prod.
1869.	May 1,	Received	2000	× 15=	30000
"	" 16,	Deliv'd	1000		
		Bal.	1000	× 10=	10000
"	" 26,	Received	4000		
		Bal.	5000	× 5=	25000
"	June 1,	Deliv'd	2000		
		Bal.	3000	× 11=	33000
"	" 12,	Deliv'd	2200		
		Bal.	800	× 20=	16000
"	July 2,	Deliv'd	800		3,0)11400,0
					3800

Chargeable for 1 mo. 3800 bbls. a 6=$228, the ans.

DOMESTIC EXCHANGE.

Domestic Exchange is the act of remitting bills to places in the same country, and when the drawer and drawee both live in different countries it is called a foriegn bill of exchange. By this means debts are discharged more conveniently than by cash remittances payable at sight or after sight.

At Sight, means at time of its presentation to the person ordered to pay.

After Sight, is requiring payment to be made at the time specified in the draft or bill.

Drawer, one who draws a bill.

Drawee, one on whom a bill is drawn.

Payee, one to whom a note is payable.

Rule.—For Sight Drafts, if at premium, multiply the face of draft by the increased rate per cent., and if at a discount, multiply the face of the draft by the decreased rate per cent.

For Draft payable after sight, find the interest of $1 where the draft is purchased; then add the rate of exchange if at a premium, or substract when at a discount; and multiply the face of the draft by this result.

Ex. 1. A merchant in New York wishes to remit

to New Orleans $5742.50, exchange 1 1-2 per cent. below par; what must he pay for a bill?

Ans. $5656.36.

100	per cent. is written thus,			1.00
1 1-2	"	"	"	.01.5
98 1-2	"	and		.98.5

Deduct 1 1-2 pr. ct. makes

5742.50
.985

5656.36.250

Point off the same decimals in the product as Ex. 2 in Insurance, page 11.

Ex. 2. A merchant at Chicago wishes to pay a bill at sight in New York amounting to $4582, and finds that exchange is 1 1-4 per cent. premium; what must he pay for a bill?

100	per cent. is written thus,			1.00
1 1-4	"	"	"	.01.25
101 1-4	"	and	"	1.01.25

Add 1 1-4 per cent. makes

4582.00
1.01.25

4639.27.5.00

Point off in the product the same number of decimals as in Ex. 4 in Banking, page 58.

Ex. 3. What will be the cost in Syracuse, N. Y., of a draft on Albany for $800, payable at sight, exchange being 3-4 per cent premium? Ans. $806.

100 per cent is written thus, 1.00
00 3-4 " " " " 00.75

100 3-4 " and " " 1.00.75

Add 3-4 per cent. makes

1st method.
1.00.75
800

$806,0000 Ans.

2d method.
4.2)800

400
200

600
800

$806.00

What costs for a bill on Buffalo for $550, at 5-8 of 1 per cent. discount ?

550
.01

8)5.50 at 1 per cent.

0.6875
5

$3.43.75

550.
3.4375

$546.56.25

Ex. 4. At sight pay to the order of Wm. F. Smith fifty-five hundred eighty-eight dollars, value received, and charge the same to GEO. R. CHAPMAN.

SUGGESTION.—Since Exchange is 2 per cent. premium, the draft is worth the amount stated in it and 2 per cent besides

We therefore find 2 per cent. of $5588.

5588.00
.02

111.76.00
5588.

$5699.76 ans.

Point off in the product the same as as Ex. 8 in Percentage, page 3.

Ex. 5. A merchant in St. Paul orders goods from New York amounting to $650, which amount he remits by draft, exchange being at a premium of 2 3-4 per cent; if he pays $27.64 freight, what will the goods cost him in St. Paul? Ans.$ 695.51.5.

```
100    per cent is written thus, 1.00
  2 3-4  "     "     "     "     .02.75
102 3-4  "    and    "     "    1.02.75

Add 2 3-4
per cent.          1.0275
makes                650.00
                  51375000
                 61650
                 667.87.5000
We add freight,   27.64
                 $695.51.5
```

Point off in the product the same decimals as Ex. 3, 1st method, 2d product, page 36.

Ex. 6. What will be the cost in Kalamazoo, Mich., of a draft on Hartford, Conn., for $600, payable in 60 days after sight, including grace, exchange at a premium of 2 per cent.? Ans. 604.65.

```
2)100                          100
    1.0                         01.225
  1000                          98775
    50                      Add  2 per ct. prem.
6)1.050 int. at 6 per cent.     1.00775
   175                               600
  01.225 is 1 cent, 2 mills      $604.55.000 ans.
and 25 hundredths of a mill.
```

Interest is 7 per cent in Michigan.

Point off in the product the same number decimals as in Ex. 4, 1st method, Commission and Brokerage, page 36.

Ex. 7. What will be the cost in Boston of a draft on Cincinnati for $560, payable in 30 days after sight, allowing grace? what is the cost of the above draft, exchange at a premium of 3 per cent.?

Ans. 573.72.

```
2)100
    5
 ----
  500
   50
 ----
5.5.0
```

5.5.0 int. is 5 mills and 5 tenths of a mill at 6 per cent.

```
          100
          .0055
          -----
          .9945
Add       3 per cent. rate of exchange.
          -----
        1.0245
           560.00
        ----------
         61470000
        51225
        ----------
        573.720000
```

Point off in the product the same as Ex. 5, p. 71.

Ex. 8. A flour dealer in Worcester receiving from his agent in Milwaukee 350 bbls. flour at $5.25 per bbl., in payment for which he remits a draft on Milwakee at 2 3-4 per cent. discount.

The freight on his flour cost $70; what must he sell it per barrel to gain $100?

```
100    per cent is written thus, 1.00
  2 3-4  "      "      "      "   .02.75
------------------------------------------
 97 1-4  "     and     "      "   .9725

Deduct                 350
2 3-4 per ct.          5.25
makes               -------
                    1837.50
                      .9725
                   --------
               1786.96.8750
                 70.00       freight.
                100.00       gain.
               ---------
          350)1956.96
              -------
               $5.59 Ans.
```

Point off in the 1st product the same as Ex. 3, 1st product, 1st method, page 54, and in the 2d product the same as Ex. 5, page 17, and in the quotient the same as Ex. 1, page 38.

Ex. 9. A merchant in Boston purchased a draft on St. Louis for $400, payable in 30 days after sight; what did it cost him, allowing grace, exchange 1 1-2 per cent discount? Ans. $391.80

```
 2)100                          100
     5                          00.55
  ----                          -----
   500                          .9945
    50                 Deduct    15  =1 1-2 rate of ex.
  ----                          -----
   5.5.0 int., the same         .9795
as Ex. 7, page 72.                400
                              --------
                             $391.80.00
```

Point off in the product the same as in Ex. 5, p. 71.

To find the Face of a Draft a given sum will purchase:

RULE.—The same as in The Proceeds of a note to Find the Face, page 60.

Ex. 10. What draft may be purchased for $487.20, exchange at 1 1-2 per cent. premium?

	100	per cent is written thus,				1 00
	1 1-2	"	"	"	"	01.5
Add 1 1-2 per ct. makes	101 1-2	"	and	"	"	1.01.5

1.015)487.200

$480 ans.

Annex a cipher to the right of the dividend to make the same number decimals equal to those in the divisor. See rule and example, page 40.

Ex. 12. What draft may be purchased for $$158.40, exchange at 1 per cent. discount?

100
1
—
99

99)158.40

$160 ans.

EXCHANGE WITH ENGLAND.

A Foreign Bill of Exchange may be defined a written order; the drawer and drawee live in different countries.

England and France are the principal countries with which United States have exchanges.

In Great Britain accounts are kept in Pounds, Shillings, Pence and Farthings.

The Pound Sterling, or Sovereign, is received at the custom house in payment of duties at	$4.84
The exchange is	4.44 4-9
To which we add 9 per cent. premium,	40
	$4.84 4-9

To change English Sterling money to U. S. money, divide by 9-40 and the quotient will be dollars. To change dollars into English Sterling money, multiply the dollars by 9-40 and the product will be Pounds Sterling.

To find the decimal of a Pound Sterling, multiply the decimal by as many of the next lower denomination as makes one of the given denomination. Point off from the product as many decimal places as are in the given decimal. Proceed thus to the lowest denomination; the figures on the left of the points are the value of the decimal.

Ex. 1. What must a merchant in Boston pay in dollars and cents for a bill of £8765 15s. 6d. on Liverpool, the premium 9 per cent.?

15.6 1 penny is 1-240 of a pound.
12 pence a shilling.

186

240)18600 two ciphers annexed. See rule in Percent-(age, p. 1.)

775 reduced to a decimal.

100 per cent. is written thus, 1.00
9 " " " " .09

109 " and " " 1.09

Add 9 per ct. makes

8765,775
40

9)350631,000

38959
1.09

$42465.31 ans.

Ex. 2. Henry Kinsman of New York, has consigned a cargo of coal, valued at £18000, to Geo. W.

Warren, of Liverpool. Wm. F. Smith, of New York, is about importing flour; has purchased of H. Kinsman a bill of exchange at 7 per cent. premium for the value of the above coal; what should be paid for the above bill?

```
 80000          18000
   .07             40
-------        -------
5600.00        9)720000
               --------
                 80000
                  5600
               --------
               $856.00
```

Ex. 3. J. M. Jones of Boston, has the net proceeds of a consignment amounting to $7000, which he is desirous of remitting to his principal in London. For what amount in Sterling money must he purchase a bill—exchange, 9 per cent.? Ans. £1444 19s.

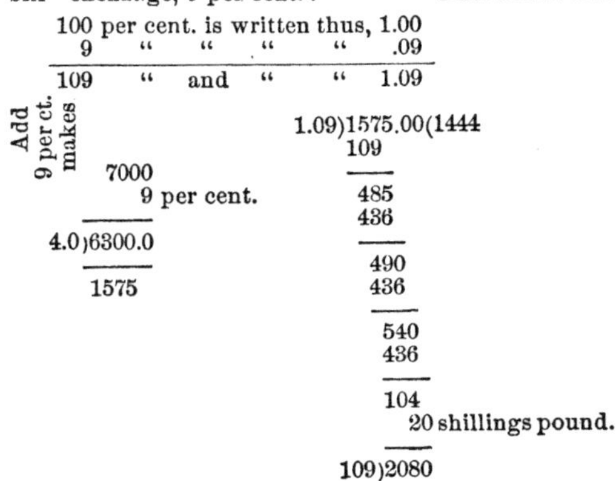

```
  100 per cent. is written thus, 1.00
    9   "    "    "    "    .09
  ------------------------------------
  109   "   and   "    "   1.09

Add 9 per ct. makes      1.09)1575.00(1444
                              109
       7000                   ---
          9 per cent.         485
  --------                    436
  4.0)6300.0                  ---
  --------                    490
     1575                     436
                              ---
                               540
                               436
                               ---
                               104
                                20 shillings pound.
                              ----
                         109)2080
                              ----
                               19s.
```

Ex. 4. What is the par value in Federal money of £248 16s. 6d.? Ans. $1105.88.8

```
16 6          240)19800        248,825
12            ---------             40
---              825          ---------
198                           9)9953,000
                              ----------
                                1105.888
```

Ex. 5. Geo. R. Chapman, of Vermont, wishes to purchase a bill of £18761 10s. on Liverpool—premium 1-2 per cent.; what will be the cost of the bill?

Ans. £20356 4s. 6 1-4d.

```
100 per cent. is written thus, 1.00          18761.50
  8 1-2  "      "     "     "   .08.5          1.085
----------------------------------------------------
108 1-2  "     and    "     "   1.085     20356.22750

And 8 1-2 per ct. makes

          240)12000(50        227 decimal of a pound.
              1200             20 shillings a pound.
     10       ----            ----
     12         0            4,540
     ---                       12 pence a shilling.
     240                      ----
                             6,480
                                4 farthings a penny.
                              ----
                             1,920
```

Ex. 6. How many dollars, cents and mills in £29 4s.?

```
 4.84 pound sterling.        24 1-5 is 1 shilling.
 29                           4 shillings.
------                       --
140,36                       96
  96.8                       0.8
------                       ---
$141.32.8 ans.               96.8
```

EXCHANGE WITH FRANCE.

Accounts in France are kept in francs and centimes. The mercantile value of some of the principal coins of France in United States currency is as follows:

The Double Napoleon or Louis, 40 francs, -	7.40
The franc, 100 centimes, is - - - -	18.6

Ex. 1. If Chas. Chapman, of Boston, should remit to Paris 18400 francs, exchange 1 1-2 per cent. premium, what will be the cost of his bill in U. S. currency?

Ans. $34737.36.

100 per cent. is written thus, 1.0
1 1-2 " " " " .01.5
101 1-2 " " " " 1.015

Add 1 1-2 per ct. makes

184000
.186=1 franc.
34224,000
1.015
34737.36.0000

Ex. 2. What must a merchant in New York pay for a bill on Havre for 7000 francs—exchange 5 francs, 8 centimes for a dollar?

5 8-100
100
508

7000
100
)700000
$13779.52

Ex. 3. What is the value of a bill on Paris for 44000 francs—exchange 2 per cent. below par?

100 per cent. is written thus, 1.00
2 " " " " .02

98 " and " " .98

Deduct 2 per ct. makes

44000
.186=18 cents, 6 mills.

8184,000
.98

$8020.32.000

Ex. 4. Henry A. Warren remits to Paris $12350—the exchange 5 francs, 10 centimes for a dollar—required the amount in francs?

5 10-100
100

510 510×12350=1.00)62985.00
100 100 1 62985 francs, ans.

DISCOUNT ON BILLS AND INVOICES.

Merchants and traders deduct a percentage from Invoices and Bills of goods sold for ready pay.

RULE.—Reckoned the same as Interest.

Ex. 1. A bought a bill of goods amounting to $962, on 6 mos. credit, but B offers to deduct 15 per cent. for ready pay; what is the amount to be deducted? 96200 Ans. $144.30
.15

144.30.00

Ex. 2. A sells B an invoice of goods for $1500, and

allows him 3 per cent. for ready pay; what amount must A receive?

1500.00	1500 00
.03	45.00
45.00.00	$1455.00

Ex. 3. A sells B 100 bbls. flour at $6 per bbl., and allows him 2 per cent. off for cash; what amount must A receive?

100	per cent.	is written	thus,		1.00
2	"	"	"	"	.02
98	"	and	"	"	.98

Deduct 2 per ct. makes	1st method.	2d method.
	100	100
	6.00	6.00
600.00	600.00	600.00
.02	.98	1200
12.0000	$588.0000	$588.00

Point off from the first product the same as Ex. 11, in Loss and Gain, p 22, and in the 2d product the same as Ex. 10 in Percentage, p. 4.

Ex. 4. If the gross price of one pair of shoes cost 1.75, what is the net cash, less 17 and 6 per cent.?

No. 1.	No. 2.	
1.75	1.75	
.17 per cent.	2975	145
		087
.29.7.5	1.45.25	
	.06	$1.36.3 ans.
	.08.7.150	

No. 1 we point off the same number decimals as in Ex. 8, in Percentage, p. 3. The product of 2 the same as Ex. 4, pages 58–9. We prefix one cipher to the product of No. 2 to make the number of decimal places equal to those in the factors.

Ex. 5. A sold goods, $3120, to be paid one-half in 3 months, and the other half in 6 months; what must be discounted for present payment?

```
                       100
                        1.5
 2)3120               ------
   ----               .01.500 interest 1 1-2 cents.
   1560               100      on $1, 3 months.
                     -----
                     1,015)1560,000
                            ---------
                             1536,945

 100
  3.0
 -----                                3120
.03.0.00=3 cts. int. on $1, 6 mos.    1536,945
 100                                  --------
 ---                                  1583,055
1.03)1560.00                          1514,563
    --------                          --------
     1514,563                         $68,492 ans.
```

Ex. 6. A sells B an invoice of marble to the amount of $4059.63; what will be the amount of his bill, less 1-6 and 3 per cent?

```
6)4059.63               4059.63      3383.03
  -------                676.60       101.49
  676.60=1-6 the amt.   -------      -------
                        3383.03     $3281.54 ans.
                            .03
                        -------
                        101.49.09
```

Point off in the product the same as Ex. 8, in Percentage, p. 3.

Ex. 7. A receives $10.50 for his work; what will he receive, discounting 15 per cent? Ans. 8.92.5.

```
        100 per cent. is written thus, 1.00
         15   "    "    "    "      .15
        ----------------------------------
         85   "   and   "    "      .85

Deduct                  10.50
15 per ct.                .85
make                    -----
                      8.92.50
```

See p. 3, Ex. 8, for pointing off the decimals from the product.

Ex. 8. A merchant bought an invoice of goods amounting to $43.60, on 30 days credit; what will he pay on his bill if paid within 30 days at 1 per cent. discount?

```
                                 43.60          43.60
1 per cent. is written thus,      .01             43
                                 -----          ------
                                 .43.60         $43.17 ans.
```

Ex. 9. What will 29 feet of belting come to at 82 cents per foot, less 25 and 8 per cent.?

```
        No. 1.
         29
         .82
        -----                     No. 2.
     4)23.78                      17,835
        5.945=1-4                   .08 per cent.
       -------                    -------
       17.835                     1.42.680
        1.426
       -------
     $16.40.9 ans.
```

See Ex. 11, p. 17 for pointing off the decimals from 1st product No. 1, and in No. 2 see Ex. 2, p. 11.

WOOD PER CORD.

Multiply the length by the height and divide the product by 32* (if the wood is 4 feet long) the quotient will be the number of cords.

If there are inches, reduce them to a decimal by the following

RULE.—Divide the numerator by the denominator; annex as many ciphers to the numerator as may be necessary. Point off as many places in the quotient as there were ciphers annexed to the numerator.

Ex. 1. How many cords of 4 foot wood are there in a pile of wood 32 feet long and 4 feet 2 inches, or 2-12 high?

Ans. 4 cords, 17 hundredths.

12)200
.17 reduced to a decimal.

4.17=4 ft. 2 in. high, nearly.
32 feet long.
32)133.44
4.17

Ex. 2. How many cords are there in a pile of wood 4 feet wide, 12 feet long and 4 feet high?

12
4
32)48
1 1-2 cords the ans.

Ex. 3. What is the cost and how many cords are there in 6 piles of wood measuring as follows:

Ans. $565.16. $66\frac{95}{100}$ cords.

*Divide by 32, because it takes so many superficial feet to make a cord.

1 pile wood 144 1-2 feet long, 10 feet 2 inches (or 2-12) high=45.923 cords at $9= 413.30

1 pile 61 ft. long, 6 ft. high=11.43 cords at $7.25= 82.86

1 do 39 ft. long, 4 1-2 ft. high=5·48 cords at $8.00= 43.84

1 do 4 ft. long, 4 ft. high=0.50 cords at $7.50= 3.75

1 do 8 ft. long, 4 ft. hight=1.00 cord at $7= 7.00

1 do 14 ft. long, 6 ft. high=2.62 cords at $5.50= 14.41

12)200 $565.16

.17 reduced to a decimal.

144.5=144 1-2 ft long.
10.17

32) 1469.565

45.923
9.00 per cord.

413.30700

39 ft. long.
4.5=4 1-2 ft. high.

32)175.5

5.48
8.00 per cord.

$43.84.00

61 ft. long.
6 ft high.

32)366

11.43
7.25 per cord.

$82.86.75

8 ft. long
4 ft. high.

32)32

1
7.00 per cord.

$7.00

4 ft. long.
4 ft. high.

32)16.00

0.50
7.50 per cord.

$3.75.00

14 ft. long.
6 ft. high.

32)84

2.62
5.50 per cord.

$14.41.00

45.92
11.43
5.48
0.50
1.00
2.62

66.95=66)
(cords, 95-100

CUSTOM HOUSE BUSINESS.

Duties are sums of money assessed by government upon imported goods.

A Specific Duty is a certain sum per ton, hundred weight, pound, hogshead, gallon, square yard, etc.

An Advalorem duty is a certain per cent. on the cost of the goods in the country from which they are imported.

A Port of Entry is a port designated by law where goods from a foreign country may be landed.

In the custom house weight and guage of goods, certain deductions are made for the hogshead, box, cask, etc., containing the goods, also for leakage, breakage, etc.

These deductions must be made before the Specific Duties are imposed.

Gross Weight is the whole weight of the goods, together with that of the casks, bags and boxes which contain them.

Draft or Tret is allowance made for waste which is to be deducted from the gross weight, and is as follows:

On	112 lbs.			it is	1 lb.
From	112	"	to 336	"	20	"
"	336	"	to 1120	"	4	"
"	1120	"	to 2016	"	7	"
Above	2026	"	any wt.	"	9	"

Consequently 9 lbs. is the greatest draft allowed.

Tare is allowance made for waste for the weight of the hogshead, box, bag, cask, etc.

It is to be deducted from the remainder of any weight or measure after the Draft or Tret has been allowed.

Net Weight is the weight of the goods after deducting the weight of the box, bale, etc., making all other allowances, customary Tare or liquors in casks is sometimes allowed on the supposition that the cask is not full, or what is called its *actual wants* and then an allowance of 5 per cent. for leakage, (a tare of 10 per cent. is allowed on porter, ale or beer in bottles, for breakage,) and 5 per cent. on all other liquors in bottles.

SPECIFIC DUTIES.

RULE.—First, deduct from the given quantity of goods the legal allowance for draft, tare, breakage or leakage.

Second, Multiply the remainder by the duty on a unit of the weight or measure of the goods, and the product will be the duty required.

Ex. 1. What is the specific duty on 120 barrels of figs, each weighing 84 lbs., gross tare, on the whole 680 lbs., at 4 cts per pound? Ans. $376.

```
  120
   84
-----
10080
  680 lbs. tare.
-----
 9400
  .04 cts. per pound.
-----
37600
```

Ex. 2. What is the duty on 396 bottles of Porter at 6 cents per bottle, allowing 10 per cent. for breakage? Ans. $256.62.

No. 1.

```
 396
  12
----
4752
 475
----
4277
 .06 cents per bottle.
------
256.62
```

No. 2.

```
4752
  10 per ct. for breakage.
------
475.20
```

In the last product of No. 1 we point off from the product the same as Ex. 11, p. 17, and in the product of Ex. 2, same as No. 9 p. 4.

Ex. 3. What is the duty on 6 bbls. of Spanish tobacco, the first weighing 174 lbs. gross, the second 154 lbs. gross, the third 122 lbs. gross, the fourth 140 lbs. gross, the fifth 50 lbs. gross, the sixth 112 lbs. gross, at 6 cts. per pound, the customary allowance being made for draft and 16 lbs. per barrel? Ans. $29.12.

```
16 lbs. per bbl. for tare.
 6 bbls.
--
96 lbs. tare.
```

```
174
154
122
140
 50
112
---
752
  4 lbs. draft.
---
748
 96 lbs. tare.
---
652 net weight.
.06 cts. per pound.
------
$39.12
```

Point off from the product the same as Ex. 2, last product of No. 1.

Ex. 4. What is the duty on 6 pipes of Port Wine, gross guage as follows: No. 1, 162 gals., No. 2, 142 gals., No. 3, 118 gals., No. 4, 144 gals., No. 5, 158 gals., No. 6, 110 gals.; wants of each pipe 4 gals., duty 15 cts. per gallon, allowing 2 per cent. waste.

```
  834       6 pipes.                162
  .02       4 gals. each pipe.      142
 -----      -                       118
16.68      24 gals.                 144
                                    158
                                    110
                                    ---
                                    834
                                     24
                                    ---
                                    810
                                     16.68
                                    ------
                                    793.32
                                         15 cts. per gal.
                                  ---------
                                  118.99.80
```

Ex. 5. What is net* weight of 6 boxes sugar weighing as follows, tare 15 per cent.: $2480\frac{10}{100}$ lbs.

```
No. 1.              No. 2.
 29.6                456
  .15                480
------               520
441.90               480
                     470
                     540
                    ----
                    2946 gross weight.
            6×4=      24 draft.
                    ----
                    2922
                     441.90
                    -------
                    2480.00 net weight.
```

*See p. 86 explaining net weight.

See Ex. 9. p. 4, for pointing off from the product of No. 1.

Ex. 6. What is the net weight of a hogshead of sugar weighing, gross, 1250 lbs., tare 12 per cent.?

```
    1243          1250
     .12             7 lbs. draft.
  ------          ----
  149.16          1243
                   149
                  ----
                  1094 lbs., ans.
```

Point off from the product the same as Ex. 5, No. 1.

ADVALOREM DUTIES.

Advalorem duties are estimated upon the actual cost of the goods. It is plain that they are found by simply multiplying the cost of the goods by the given per cent.

Note.—In advalorem duties no deductions of any kind are to be made.

Ex. 1. What is the advalorem duty at 28 per cent. on 30 yards of English broadcloth which cost $4.75 per yard? Ans. $39.90.

```
    4.75
      30
  ------
  142.50
     .28 per cent. duty.
  ------
39.90.00
```

Ex. 2. What is the advalorem duty at 20 per cent. on a piece of Turkey carpeting, containing 145 yards, and cost $1.88 per yard; what is the duty on one yard? For how much must it be sold, per yard, to gain 25 per cent. on the cost and duty? Ans. $2.82.

```
               100 per cent. is written thus, 1.00
                25   "     "     "     "   .25
               ---------------------------------
               125   "    and    "     "  1.25

  Add
  25 per ct.     188            145 yds.
  makes           37.6           1.88
                 ------         ------
                 2.25.6         272.60
                  1.25             .20
                 -------        -------
                 2.82.000       $54.52.00 duty advalorem.
                          cost 1.88 per yard.
                                 20 per cent. duty.
                                -----
```

Duty on 1 yd. is 37.60=37 cts. 6 mills.

Ex. 3. What is the duty on an invoice of goods which cost in London £964 sterling at 44 per cent. advalorem?

The pound sterling being $4.84. Ans. $2052.93.

```
                 964
                 4.84
             ---------
             4665.76 1st product.
                 .44 per cent. duty.
          ------------
          $2052.93.44 2d product.
```

FEDERAL MONEY.

Federal currency of the United States was established by Congress in 1786. Its denominations are eagles, dollars, dimes, cents and mills.

To read Federal money, call all the figures on the left of the decimal point dollars, the first two figures on the right of the point cents, the third figure mills, the other places on the right decimals of a mill.

Thus, $4.26.2.32 is read 4 dollars, 26 cents, 2 mills and 32 hundredths of a mill.

ASSESSMENT OF TAXES.

Assessment of taxes is a sum imposed on an individual for a public purpose.

A pole tax is a specific sum assessed on male citizens above 21 years of age; each person assessed is called a poll.

Taxes are usually assessed either on the person or property of the citizens, and sometimes on both.

Property is of two kinds, personal property and real estate.

Personal is *movable* property, such as money, notes, cattle, furniture, etc.

Real estate is *unmovable* property, such as land, houses, stores, etc.

An Inventory is a list of articles.

RULE.—First find the amount of tax on all the polls, if any, at the given rate, and substract this sum from the whole tax to be assessed. Then divide the remainder by the whole amount of taxable property in the State, county or town, etc.; the quotient will be the per cent., or tax on one dollar.

Second, multiply the amount of each man's property by the tax on one dollar, and the product will be the tax on his property.

Third, add each man's poll tax to the tax he pays on his property, and the amount will be the whole tax.

Proof.—Where a tax bill is made out, add together the taxes of all individuals in the town, district, etc., and if the amount is equal to the whole tax assessed, the work is right.

TABLE.

$1	pays	.03	$ 20	pays	.60	$ 300	pays	$9.00
2	"	.06	30	"	.90	400	"	12.00
3	"	.09	40	"	1.20	500	"	15.00
4	"	.12	50	"	1.50	600	"	18.00
5	"	.15	60	"	1.80	700	"	21.00
6	"	.18	70	"	2.10	800	"	24.00
7	"	.21	80	"	2.40	900	"	27.00
8	"	.24	90	"	2.70	1000	"	30.00
9	"	.27	100	"	3.00	2000	"	60.00
10	"	.30	200	"	6.00	4000	"	120.00

In making out a tax list by the preceeding table containing the taxes on 1, 2, 3, etc., to 10 dollars, then 20, 30, to 200 dollars, then 300, 400, to 4000 dollars; then knowing the inventory of any individual it is easy to find the tax on his property.

Ex. 1. In the above assessment, what is a man's tax who is rated at $2256, and pays for 3 polls at 50 cents each.

Operation.

$2000	pays	60.00
$200	"	6.00
$50	"	1.50
$6	"	.18
3 polls		1.50
		$69.18 ans.

Now if we add together the tax paid on each of these sums as found by the preceeding table the amount will be the tax on $2256.

Ex. 2. A township composed of 16 citizens, levies a tax of $5700. The town contains 30 polls which are assessed 50 cts. each, and its taxable property is inventoried at $199,500; what amount of tax must be raised to pay the debt and 5 per cent. commission for collection, and what is the tax on a dollar.

Ans., the sum to be raised is $6,000,
and the tax is 3c. on the dollar.

```
                 100 per cent. is written thus, 1.00
                   5   "     "      "      "    .05
                 ----------------------------------
                  95   "   and      "      "    .95)5700.00
Deduct                                               -------
5 per ct.          50           6000                 6000
makes              30             15
                 -----          ----
                 15.00          5985                       (p. 1.
                   Multiply by   100 see rule in Percentage,)
                                ----
                        199500)5985.00
                               -------
                                  0.03
```

Point of from the quotient the same as Ex. 2, p. 26.

No. 3. A tax of $29505 is levied on a certain county whose property is valued at $1,125,750, and which has a list of 11,650 polls, which is assessed at 60 cents apiece; what per cent is the tax and what is the amount of A's tax who pays for 6 polls and has property valued at $5625?

```
 11,650 polls.
     60 cts each.                 29,505.00
----------                         6990.00
$6990.00   Whole sum of           ---------
   taxable prop'ty, 1,125,750)22515.00
                              ---------
                                  0.02 cts. on the dol.
                    5625
                     .02
  .60              ------
   6 polls.       112.50
  ----              3.60
  3.60             ------
                 $116.10 Amount of A's tax.
```

Ex. 4. *If* a tax of $750 is assessed on a district to build a new school house, the property of the district is valued at $15,000; what is the tax on the dollar, and what is a man's tax whose property is valued at at $1150?

```
     1150           15000)750.00
      .05                 ------
    -----                 0.05 cts. on the dol.
   $57.50
```

Ex. 5. If my town tax is 25 per cent. on my grand list, and I pay 50 cts. tax, what is the amount of my grand list.

```
        50
       100
       ---
    25)5000
       ----
 Ans. $2.00
```

Ex. 6. If my grand list is $185.72, what is the amount of the following taxes made upon the grand list of 1870?

Town tax of 76 per cent. is - . -	141.14
Union tax of 15 " " - - - - -	27.85
Dis. No. 19, 30 " " - - -	55.71
Village corporation 18 per cent. is - -	33.42
Amount of tax is - - - - -	$258.12

185.72	185.72	185.72	185.72
.76	15.	.30	.18
141.14.72	27.85.80	55.71.60	33.42.96

Point off from the prodncts the same as Ex. 8. p. 3.

Ex. 7. What is the amount of my tax on $5675 at 7-10ths of 1 per cent? Ans. $39.72.5.

```
 5675.00
      .01=1 per cent.
 --------------------
 56.75.00 at I per cent.
      .7= 7-10 "
 --------------------
 39.72.500 at 7-10 per cent.
```

Point off in the 1st product the same as Ex. 6, 2d product, page 38; and in the 2d product the same as Ex. 2, page 11.

Ex. 8. The city of Chicago levied a tax of 1-8th of 1 mill on all real and personal property; what is the amount of tax on $7500? Ans. 93 cts., 7 mills.

```
   7500.00
      .001=1 mill.
  ---------
8)7.50.000
  ---------
  0.93.750
```

The decimal places in the dividend exceed those in the divisor by 5, hence we point off 5 decimals from the quotient; all the figures at the right of the point will be cents and mills.

Ex. 9. A tax of $3900 is to be assessed on the town of Rutland. The real estate is valued at $840,000, and the personal property at $210,000; and there are 500 polls, each of which is taxed $1.50;—what assessment on $1?

```
 840000                 1.50
 210000                  500
-------              -------
1050000              $750.00
              3900
               750
              ----
              3150
               100
             -----
    1050000)3150.000
            --------
             0.003 mills ans.
```

Ex. 10. What is A's tax, whose real estate is valued at $7436, and his personal property at $3215, and who pays for 6 polls?

```
  1.50              7436
     6              3215
 ------            ------
 9.00.0            10651
                     .003
                   ------
                   31.953
                     9.00
                   ------
                  $40.95.3
```

Ex. 11. A certain district paid $130 for teacher's salary, $34 for board, $19.42 for fuel and $2.58 for repairs; the district drew $30 public money, and the whole number of days' attendance was 2400; what was the rate per day? and how much was A's tax, who sent 115 days?

```
        130.  teacher's salary.
         34.     "      board.
         19.42 for fuel.
          2.58  " repairs.
        ------
        186.00
         30
        ------
        156.00                115 days.
             100              .065
        --------             ------
   2400)1560.000             $7.475 A's tax.
        --------
          .06.5 ans.
```

To find the percentage see Rule, page 1. The decimal places in the dividend exceed those in the divosor by three, counting the cipher annexed; hence we point off three decimal places in the quotient; all the figures at the right of the point will be cents and mills.

ASSESSMENT OF TAXES TO RAISE A GIVEN NET AMOUNT.

RULE.—Subtract the given per cent. from $1, and the remainder will be the net* value of $1 assessment.

Divide the net amount to be raised by the net value of $1 assessment, and. the quotient will be the sum assessed.

Ex. 1. Allowing 5 per cent. for collection, what sum must be had to raise $12425, net?

```
 100                    .95)12425.00
   5                        ___________
 ___                        $13078.947 ans.
 .95 net value of $1.
```

SIMPLE PROPORTION.

Simple Proportion is called the Golden Rule from its excellent performance in arithmetic or in other points of mathematical learning.

And it is called the Rule of Three, because from three numbers given proposed, or known, we find out a fourth number required, or unknown, which leaves such proportion to the third as the second does to the first number.

To find out a fourth proportional to three given numbers placed thus,

```
4 : 8 :: 6
      6
     ___
    4)48
     ___
      12
```

*Clear of all charges.

And multiply the second and third numbers together and divide by the first; the quotient is 12, which leaves the same proportion to 6 that 8 does to 4.

Ex. 1. If 24 lbs. butter cost $5.29, what is the price of 3 lbs?

$$\begin{array}{r} 24 : 3 :: 529 \\ 3 \\ \hline 24)1587 \\ \hline 66\ 1\text{-}8 \text{ cents the ans.} \end{array}$$

RULE WITH EXAMPLE.—In this question there are two things mentioned, butter and money, is the answer to the question to be given in butter or money. You see at once it is to be given in money.

Put down the money, $5.29, for the third term. Having done this, you have now to consider where you are to place the 24 lbs. and 3 lbs.

Read over the question and you will see that the answer must be less than the third term; for 3 lbs. will not cost as much as 24 lbs.

If, then, the answer is to be less, put the less number for the second term, and the greater for the first.

In all questions let the third term be the same as the answer, and if the answer is to be greater than the third term, put the greater second; if it is to be less, put the less second.

Ex. 2. If 11 yards makes 2 rods, how many rods are there in 1617 yards?

$$\begin{array}{r} 11 : 1617 :: 2 \\ 2 \\ \hline 11)3234 \\ \hline 294 \text{ rods.} \end{array}$$

The answer to the question is to be given in rods; place rods for the third term.

The answer 2 is to be more than the third term, therefore place the greater number for the second term, and the less number for the first.

Ex. 3. If 21 barrels of flour cost $180, what will 112 barrels cost at the same rate.

```
21 : 112 :: 180
             112
          ------
       21)20160
          ------
          $9.60 ans.
```

Ex. 4. What will 136 feet of wood come to at $4 per cord?

```
128 : 136 :: 4.00
             4.00
            -----
      128)544.00
            ------
            $4.25 ans.
```

128 cubic feet in a cord of wood.

Ex. 5. What will 22 inches of beaver cloth come to at $2.50 per yard?

```
36 : 22 :: 2.50
             22
          -----
       36)55.00
          -------
          $1.52 7-9 ans.
```

1 yd. is 36 inces long.

COMPOUND PROPORTION.

When, in order to find a fourth proportional, several circumstances require to be considered, it is called Compound Proportion.

Ex. 1. If 14 horses eat 56 bushels of oats in 16 days, how many bushels will be required for 20 horses 24 days?

```
 14 : 20 :: 56
 16   24
 ---------
224  480
      56
    -----
224)26880
    -----
      120 bushels ans.
```

RULE WITH EXAMPLE.—Write down for the third term that number which is of the same kind with the answer required—56 bushels.

Then take two numbers of the same kind, 14 horses and 20 horses, and consider, as in Simple Proportion, whether from the nature of the question, the greater or less is to be put in the first or second term.

Here it is obvious that the greater must be in the second term, as 20 horses will eat more than 14 horses.

Take the other two terms and proceed in the same manner. After all the terms have been put down, multiply the first terms, 14 and 16, together; do the same with the two 2d terms, 20 and 24, and that product by the third term, 56, and proceed as in Simple Proportion.

Ex. 2. If 8 men can reap 32 acres in 6 days, how many acres can 12 men reap in 15 days.

```
8m. : 12m. :: 32
6d.   15d.
----------
48    180
       32
     ----
48)5760
     ----
     120 acres ans.
```

Ex. 3. If 100 gain $6 in 12 months, what will $400 gain in 8 months?

```
  100 : 400 :: 6
   12     8
-------------
12 0   3200
          6
       -----
1200)19200
       -----
      $16.00
```

Ex. 4. If $100 gain $6 in 12 months, in what time will $400 gain $16?

```
 400 : 100 :: 12
   6    16
-----------
2400  1600
        12
      -----
2400)19200
      -----
         8 months ans.
```

Ex. 5. If 12 horses, in 5 days, draw 44 tons of stone from a qnarry, how many horses would it require to draw 132 tons in 18 days?

44 : 132 :: 12
 18   5
792 660
      12
792)7920
      10

Ex. 6. A garrison of 1500 men has provisions for 12 weeks, at the rate of 20 ounces per day to each man, how many men will the same provisions maintain for 20 weeks, allowing each man only 8 ounces per day?

8 : 20 :: 1500
20 12
160 240
    1500
160)360000
    2250 men ans.

BOARDS.

Rule.—Multiply the length in feet by the width in inches, divide the product by 12, the quotient will be the number of square feet.

Boards or plank that taper gradually, add the width of the two ends together, and half their sum, multiplied by the length, will be the number of square feet.

Ex. 1. How many square feet in a stock of 15 boards, 12 feet 8 inches long, 13 inches wide?

See rule in Duodecimals.

```
12  8'
 1  1
---------
 1  0'  8''
12  8'
---------
13  8'  8''
        15
---------
205 10'     ans.
```

Ex. 2. How many square feet are there in a board 18 feet long, 10 inches wide?

```
    18
    10
   ---
12)180 sq. in.
   ---
    15 sq. ft.
```

Ex. 3. How many square feet are there in 660 boards, each 16 feet long and 6 inches wide?

```
   16 feet long.          660 boards
    6 inches wide.          8 ft. in a board.
   --                     ----
12)96                     5280 ans.
   --
    8 ft. in each board.
```

Ex. 4. How many feet, board measure, are there in 1440 ft. of 1 1-4 inches thick?

```
4)1440
   360
  ----
  1800 ft. ans.
```

Ex. 5. What length must be out of a board 8 1-2 inches broad to contain a square foot?

RULE.—Divide 144 by the inches in breadth, and the quotient will be the length of that board that will make a foot.

```
 8 1-2        inverted             17)288
 2              2×144=288            ---     16
----          ----------            16      --
17            17       17                Ans. 17
--
 2
```

Ex. 6. How many square feet are there in a board 18 feet long, 13 inches wide at one end, and 17 inches wide at the other.

```
  13           18
  17           15
  —            ——
2)30        12)270
  —            ———
  15            22 1-2 ft. ans.
```

PLANK, JOISTS AND SCANTLINGS.

Ex. 1. How many feet, board measure, that is, one inch thick, are there in a stock of 15 inch plank, 12 ft. 8 inches long, 2 ft. wide and 2 inches thick?

```
feet. inches.
 12    8'
  2    0'
 —     -
 25    4' of 1 inch thick.
       2 inches thick.
 ——————
 50    8' of 2 inches thick.
      15 plank.
 ——————
760    0' ans.
```

Ex. 2. How many feet, board measure, are there in 9 plank, 1 foot wide, 3 inches thick, length as follows:

```
1st method.
  12
  12                2d method.
  12        3ps 3×12=12 ft long=108
  14        2ps 3×12=14  "     = 84
  14        1ps 3×12=16  "     = 48
  16        3ps 3×12=18  "     =162
  18                            ——
  18                  2d ans. 402
  18
 ——
134 linear feet.
  3 inches thick
 ——
402 feet 1st ans.
```

JOISTS AND SCANTLINGS.

Ex. 1. How many feet, board measure, are there in 20 joists, 10 feet long, 6 inches wide and 2 inches thick?

```
                     10 ft. long.
                      6 inches wide.
                     ---
                     60
                      2 inches thick.
                     ---
12 inches a foot)120
                     ---
                     10 ft. in each stick.
                     20 joists.
                     ---
                    200 ft. ans.
```

Ex. 2. How many feet, board measure, are there in 13 sticks of scantling, 13 feet long, 2×4.

```
                     13 ft. long.
                      4 inches wide,
                     ---
                     52
                      2 inches thick.
                     ---
12 inches a foot)104
                     ---
                      8     8′
                           13 sticks.
                    ----------
                    112 ft. 8′ ans.
```

TIMBER.

Ex. 1. How many feet, board measure, are there in 12 sticks of timber 12×12, lengths as follows, viz:—12, 12, 14, 16, 16, 18, 18, 22, 22, 24 and 26, and 4 sticks 12×14, lengths as follows:—16, 16, 16 and 16?

1st method.

```
   12×12
   12
   12
   14
   16
   16
   18
   18
   18
   22
   22
   24
   26
   ---
   218 linear feet.
    12 inches wide.
   ----
  2616
    12
  -----
12)31392
  -----
  2616
   896
  ----
  3512 1st ans.
```

2d method.

```
2 ts. 12×12×12=288
1 "   12×12×14×168
2 "   12×12×16=384
3 "   12×12×18=648
2 "   12×12×22=528
1 "   12×12×24=288
1 "   12×12×26=312
4 "   12×14×16=896
                ----
       2d ans. 3512

   12×14
     1
     16
     16
     16
     --
     64 linear feet.
     14 inches wide.
    ---
    896
     12 inches thick.
   -----
12)10752
   -----
    896
```

Ex. 2. How many feet, board measure, are there in 20 car sills, each 30 feet long, 5 1-2 by 10 1-2 inches, and how many feet, board measure, are there in 18 car sills, 5 by 7 inches, by 29 feet long?

Ans. to the 1st, 2887 40-100 feet.
" " 2d, 1522 44-100 feet.

30 ft. long.	29 ft long.
5.5=5 1-2 inches thick.	5 inches thick.
165.0	145
10.5=10 1-2 in. wide.	7 inches wide.
12)1732.50	12)1015
144.37	84.58
20 sills.	18 car sills.
2887.40	1522.44

Ex. 3. What is the solidity of a tapering square stick of timber, the largest end being 14 inches square, the lesser end 10 inches square, the length 40 feet ?

Rule.—Square the two ends, add together the area of the two ends, and one-half their sum, multiplied by the length, and divided by 144, will give the solid contents.

14	10	
14	10	
196	100	148
100		40 feet long.
2)296		144)5920
		41 1-9 feet ans.
148		

Ex. 4. If a piece of timber be 8 inches square, what length of it will make a foot.

Rule.—Multiply one by the other, and let the product be a divisor to 1728.

8
8
64)1728
27 inches long ans.

Ex. 5. How many cubic feet are there in a stick of timber 15 feet three inches long, 2 feet 4 inches wide, and 1 foot 8 inches thick? How many feet, board measure?

```
15  3′
 2  4′
---------
 5  1′  0″
30  6′
---------
35  7′
 1  8′
---------
23  8′  8″
35  7′
---------
59  3′  8″ ans. to the 1st.
```

```
35  7′
    20 inches thick.
---------
71  8′ ans. to the 2d.
```

Ex. 6. How many feet, board measure, ir 41 sticks of timber, 20 feet long, 10′ by 10′.

```
                    20 feet long.
                     10 inches wide.
                   ----
                   200
                     10 inches thick.
                   ----
12 inches a foot)2000
                   ------------
                    166,66=166 66-100 ft. in each stick.
                         41 sticks.
                    -------
                    6833.06
```

CARPETING ROOMS.

RULE.—Multiply the length in feet by the width in feet, and divide the product by 9, the quotient will be the number of yards if the carpet is one yard wide.

If less than one yard wide, we first find how many pieces we want, then multiply them by the width of each piece in inches the product will give the width of room in inches, and that product by the length of room in inches, and divide the product by the number square inches in a square yard, the quotient will be the number yards.

Ex. 1. How many yards of carpeting, one yard wide, will it take to cover a floor 18 feet square.

```
   18
   18
   ——
9)324
   ——
   36 yards ans.
```

Ex. 2. How many yards carpeting, one yard wide, will it take to cover a floor 16 feet 5 inches long, and 13 feet 7 inches wide ?

```
        13 ft  7′ wide.
We add   1 "  5′ to make full width of pieces.
        ————————
Making  15 "  0′ width of room.
        16 "  5′ length of room.
      ——————————
     9)246 "  3′
      ——————————
        27 y 3 sq. ft. 3 inches.
```

Ex. 3. How many yards of carpeting, 3-4 yard wide, will it take to cover a floor 18 feet long and 15 feet wide?

```
                    15 feet wide.
                    12 inches a foot.
                   ----
Car. 27 in. wide.)180
                   -----
                    6 2-3
To which we add     1-3 to make full width of piece.
                   ----
                    7 0' pieces.
Each piece,        27    inches in width.
                   --
                   189   inches width of room.
                   216     "    length     "
                  -----
          972)40824
              -----
                   42 yards ans.

  18 ft. long.      1 yd. 36 in. long.
  12 in. a ft.            27 in. wide.
  --                      --
  216                     972 square in's. in a square yd.
```

PLASTERING ROOMS.

Painting, plastering, paving and some other kinds of work are done by the square yard.

If the contents in square feet be divided by 9, the quotient, is evident, will be square yards.

Ex. 1. What will it cost to plaster a room 20 feet 6 inches long, 18 feet wide and 10 feet high at 12 1-2 cts. per square yard? Ans. $15.81

```
Overhead.                     20 1-2 feet long.
  20 1-2 feet long.           10       "   high.
  18       "  wide.           ---
 ---                          205
 369                            2 sides.
                              ---
                              410
        18 ft. wide.          360
        10 ft. high.          369
       ---                    ----
       180                   9)1139
         2 sides              -----
       ---                     126.55=126 55-100 sq. yds.
       360                       .125=12 1-2 cts.
                            ----------
                            $15.81,875
```

PAINTING ROOMS.

Solid bodies being frequently painted, it is necessary to know how to find their superficiality.

To find the superficial contents of a square or many sides, or round pillar, multiply the sum of the sides, or circumference, by the height in feet, and the product, divided by 9, will be square yards.

Proper Directions for Joiners, Painters, Glaziers, etc.

Rooms being varied in their form, take this general rule in all cases, viz :

Take a line and apply one end of it to any corner of the room ; then measure the room, going into every corner with the line till you come to the place where you first began ; then see how many feet and inches the string contains, and set down for the compass or round ; then take the height by the same method.

All kinds of work, such as cornice, mouldings, etc., are measured by running measure.

Ex. 1. A man painted the walls of a room 8 feet 2 inches in height, and 72 feet 4′ in compass, (that is, the measure of all the sides); how many square yards did he paint?

See rule in Duodecimals.

```
      72    4′
       8    2′
    ----------------
      12    0′    8″
     578    8′
    ----------------
   9)590  · 8′    8″
    ----------------
     65 yds 5 sq ft 8″
```

PAPERING ROOMS.

Ex. 1. How many yards of papering that is 30 inches wide, will hang a room that is 65 feet circuit 12 feet high, deducting 1-4 for doors and windows?

```
30 inches wide.            65
36   "   long=1 yard.      12 in a foot.
----                       ---
1080                       780
                           144 sq. inches sq. ft.
                        -------
                      4)112320
                         28080
                        ------
                 1080)84240
                        ------
                           78 yards ans.
```

FLOORING ROOMS.

There is a house containing 2 rooms, each 16 feet by 15 feet 4 inches, a hall 24 by 10 feet 6′, 3 bedrooms each 11 feet 4′ by 8 feet, a pantry 7 feet by 9 feet 6′, a kitchen 14 feet 2′ by 18 feet, and two chambers, each 16 feet by 20 feet 8′; what did the work of flooring cost at 2 cents per square foot?

```
 16  0′              24  0′              11  4′
 15  4′              10  6′               8  0′
 ----------          -----------         ------
  5  4′  0″          12  0′  0″          90  8′
240  0′             240                    3 rooms.
 ----------          -----------         ------
245  4′             252  0′             272  0′
   2 rooms.
 ------                        16  0′
490  8′                        20  8′
                               ------
            14  2′             10  8′
            18  0′            320  0′
            ------             ------
           255  0′            330  8′
                                2 rooms.
                              -------
  7  0′                       661  4′
  9  6′                       255  0′
 ----------                    66  6′
  3  0′  0″                   490  8′
 63  0′                       252  0′
 ------                       272  0′
 66  6′                      --------
                             1997  0′
                              .02     cts. per sq foot.
                             ------
                           $39.95 ans.
```

LATHING ROOMS.

Lathing is done mostly by the square yard. A bunch of lathing generally contains from 50 to 100 strips, each 4 feet long and 1 1-2 inches wide, making 50 feet, (allowing 100 strips to a bunch), will spread, after they are nailed on, about 63 feet, which makes 7 square yards.

Ex. 1. How many feet in a bunch of lathing containing 100 strips each, 4 feet long and 1 1-2 inches wide?

```
    4 feet long.
    1 1-2 inches wide.
   ---
    6 square inches.
  100 strips.
  ---
12)600 square inches.
   ---
   50 square feet ans.
```

Ex. 2. How many bunches of lathing 4 feet long 1 1-2 inches wide, (containing 100 strips), will it take to lath a room 20 feet 6 inches long, 18 feet wide and 10 feet high, allowing 63 feet to a bunch?

```
         18 feet wide.         20 1-2 feet long.
          10 feet high.         10 feet high.
         ---                   ---
         180                   205
           2 sides.              2 sides.
         ---                   ---
         360                   410 length.
                               360 width.
20 1-2 feet long.              369 overhead.
18     feet wide.             ----
---                        63)1139
369                        --------
                              18 5-63 ans.
```

GLAZIERS WORK.

RULE.—To find the dimensions of their work, multiply the height of windows by their breadth.

Glaziers are to take the depth and breadth of their work, multiply one by the other, dividing by 144. Glass being measured as boards.

If the windows are arched or have a curved form, no allowance is made by reason of the extraordinary trouble and waste of time, expense, or waste of glass, etc.

Dimensions taken from the highest part of the arch down to the bottom of the windows from the height or length which, multiplied by the breadth, the product will be the answer in feet, etc.

Ex. 1. There is a house which contains 22 windows, each window has 12 lights of 13 inch by 10 inch glass; what will the glazing work come to at 12 cents per square foot? Ans. $28.59.

13 inches long.
10 " wide.
———
130
12 lights in a window.
———
1560
22 windows.
———
144)34320
———
238,33
,12 cents per square foot.
———
$28,59,96

Point off from the last product the same as Ex. 8, p. 3.

TO MEASURE ROUND TIMBER.

RULE.—Multiply the length in inches by the square of 1-4 the girth in inches, and the product divided by 1728 will give the contents in cubic feet.

If a tree or timber is tapering, girt it about 1-3 of the way at the two ends, and divide the sum by 2 to obtain the mean girth.

Allow on account of bark, in oak 1-10 or 1-12 of the circumference; beech, ash, etc., should be less.

Ex. 1 A stick of tember is 18 feet long and 56 inches girt; how many cubic feet does it contain?

```
4) 6                 14
   —                 14
  14=1-4 the girt.   —
                     196
                     216 inches long.
                     ————
              1728)42336
                     ————
                     24 1-2 feet ans.
```

TO FIND THE BOARD MEASURE IN A LOG.

RULE.—Subtract 4 from the diameter in inches, (which is taken at the smaller end), multiply by one-half the difference, and that product by the length in feet, and divide by 8.

Ex. 1. If the diameter of a log is 16 inches, and the length is 20 feet, how many feet, board measure, does it contain?

```
    16 inches in diameter.
     4
    —
    12
     6=half the difference.
    —
    72
    20 feet long,
    ——
  8)1440
    180 feet ans.
```

SLATING AND SHINGLING ROOFS.

Slating roofs is estimated by the "square," a builder's measure, 10×10 feet. All hips and gutters are measured in length, at one foot wide, and added to the surface.

SIZE OF SLATE MANUFACTURED BY THE EAGLE SLATE COMPANY.

Size	Number of Slate in a square.	Size	Number of Slate in a square.	Size	Number of Slate in a square.	Size	Number of Slate in a square.	Size	Number of Slate in a square.	Size	Number of Slate in a square.
14×7	374	16×8	277	18×9	213	20×10	169	22×11	137	24×12	114
14×8	327	16×9	296	18×10	192	20×11	154	22×12	126	24×13	105
14×9	290	16×10	221	18×11	174	20×12	141	22×13	116	24×14	98
14×10	261			18×12	160					24×16	85
12×8	400										
12×7	457										

SLATING ROOFS.

Ex. 1. How many slates, 10×10, will it take to cover one square, (that is 100 square feet).

144 in. in a square foot.
100 square feet.

65)14400

221 7-13 slates ans.

16 inches long.
Deduct 3 inches for underlap.

2)13

6 1-2 inches to the weather.
10 " wide.

65 square inches.

Ex. 2. How many squares are there on both sides of a roof whose ridge is 30 feet long and rafters 20 feet long; how many slates, 18×10 inches, will it take to cover both sides of the roof, including the under-course?

Ridge, 30 feet long.
12 inches a foot.

Slate 18 in. long.)360

20 slates for undercourse on both sides.
2 sides.

40 slates for undercourse on both sides.

Ridge 30 feet long.
20 " long, rafters.

600
2 sides.

100)1200
12 squares, ans. to first.

192 slates to a square by the table.
 12 squares.

2204
 40 slates for undercourse.

2244 Ans. to the last.

SHINGLES FOR A SQUARE.

Ex. 3. How many shingles, 4 inches wide, and laid 3 1-2 inches to the weather, will it take to lay a square?

4
3 1-2

14 square inches.

144 inches in a square foot.
100 square feet.

14)14400

1028 4-7 shingles, ans.

PERCHES IN CELLAR WALL.

Mason's work is sometimes estimated by the perch of 16 1-2 feet in length, 1 1-2 feet in width and 1 foot in height.

A perch contains 25.75 cubic feet. If any wall be 1 1-2 feet thick, its contents, in perches, may be found by dividing its superficial contents by 16 1-2 feet, but if it be any other thickness than 1 1-2 feet, its cubic contents must be divided by 24.75 (24 3-4) to reduce it to perches.

Joiners, painters, plasterers, bricklayers and masons sometimes estimate their work by the girt; that is, the length of the wall on the outside.

Ex. 1. The side walls of a cellar are each 32 feet 6′ long, the end walls 24 feet 6′, and the whole 7 feet high and 1 1-2 feet thick; how many perches of stone are required, allowing nothing for waste, and for how many perches must the mason be paid for ?

Ans., 45 9-11 perches in the work. The mason must be paid for 48 4-11 perches.

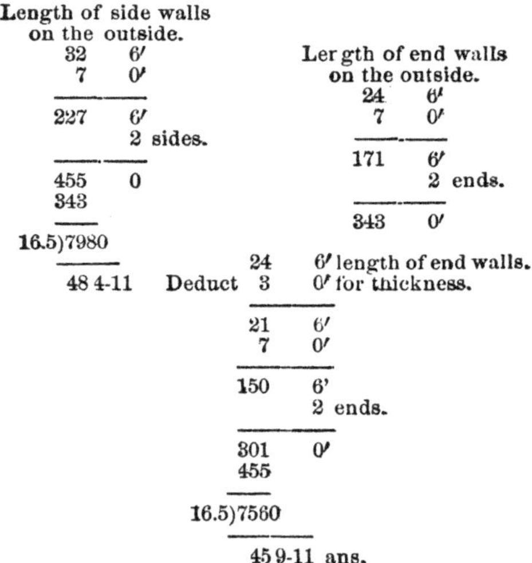

```
Length of side walls
  on the outside.
     32    6'                      Length of end walls
      7    0'                        on the outside.
   ─────────                            24    6'
    227    6'                            7    0'
           2 sides.                   ─────────
   ─────────                           171    6'
    455    0                                  2 ends.
    343                               ─────────
   ────                                343    0'
16.5)7980
   ──────                   24    6' length of end walls.
    48 4-11       Deduct     3    0' for thickness.
                          ─────────
                            21    6'
                             7    0'
                          ─────────
                           150    6'
                                  2 ends.
                          ─────────
                           301    0'
                           455
                          ────
                     16.5)7560
                          ──────
                           45 9-11 ans.
```

CUBIC YARDS IN CELLAR WALL.

Ex. 1. The side walls of a cellar are each 32 feet 6' long ; the end walls in the inside are 22 feet long, and the whole 7 feet high and two feet thick, and for how many cubic yards must the mason be paid for ?

Ans. 56 51-100 cubic yards.

32 6′ ft. long, side walls.
7 0′ ft. high.

227 6′
2 0′ feet thick.

455 0′
2 sides.

910 0′

22 0′ ft. long, end walls.
7 0′

154 0′
2 ft. thick

308 0′
2 ends.

616 0′
910 0

27 cu. ft. in a yd.)1526

56.51

How many cubic yards in the above walls, extending the length of the end walls to the center of both side walls. Ans. 58 59-100 cu. yds.

32 6′ long, side walls.
7 0′ high.

227 6′
2 0′ ft. thick.

455 0′
2 sides.

910 0
672

27)1582

58.59 cu. yds. ans.

22 0′ long, end walls.
2 0′ which is the center of side walls.

24 0′
7 0′ feet high.

168 0′
2 0′ thick.

336 0
2 ends.

672 0′

Note.—The masons are sometimes allowed to the center of the side walls.

BOARDS TO COVER A HOUSE.

Ex. 1. How many feet of boards will it take to cover the walls of a house 40 9-12 feet long, 30 1-2 feet wide, and 20 feet high, allowing 5 per cent. waste?

40 3-4 feet long.	30 1-2 feet wide.
20 feet high.	20 feet high.
815	610
2 sides.	2 sides.
1630	1220
.05 per cent.	1630
81.50	2850
	81
	2931 feet, ans.

BRICKS FOR THE WALLS OF A HOUSE.

Ex. 1. How many bricks, 8 inches long, 4 inches wide and 2 inches thick, will it take to build the walls of a house which is 80 feet long, 40 feet wide and 25 feet high, the wall to be 12 inches thick?

Ans. 159300 bricks.

40 feet wide.
Deduct 2 feet for thickness of wall.

	38
960 in. long	12 in. a foot.
300 in. high.	456 in. wide.
288000	300 in. high.
2 sides.	136800
576000	2 sides.
12 in. thick.	273600
6912000	12 inches thick.
	3283200
	6912000

1728 cu. in. in a cu. ft.)10195200

5900 cubic feet.
27 bricks in a cubic foot.

159300 bricks.

Some masons allow 23 bricks in a cu. ft. in mortar.

BRICKS FOR FLOORING ROOM.

Ex. 1. How many bricks, 9 inches long, 4 inches broad, will it take to floor a room 20 feet square?

```
 9 inches long.              20 feet long.
 4    "    broad.            20  "  broad.
 —                           ——
36)144 in. a sq. ft.         400
      4 bricks in a sq. ft.    4 bricks in a sq. ft.
                             ————
                             1600 bricks, ans.
```

BUSHELS IN BIN.

Rule.—Reduce the cubic feet to cubic inches, then divide by the number of cubic inches in a bushel, which is 2150.4; the quotient will be the answer.

Ex. 1. How many bushels of grain will a bin hold 48 feet long, 16 feet wide and 14 feet deep?

```
               48 feet long.
               16 feet wide.
              ———
              768
               14 feet deep.
             ————
            10752
             1728 cu. in. in a cu. ft.
       ——————————
2150.4)18579456.0
       ——————————
             8640
```

To find the number of bushels, heaped measure, divide the number of cubic inches by 2747.7.

GALLONS IN A BARREL.

Ex. 1. How many wine gallons in a cistern, which is 6 feet long, 5 feet wide and 4 feet deep.

Ans. 897 gals., 2 qts., 1 pt., 1 gill.

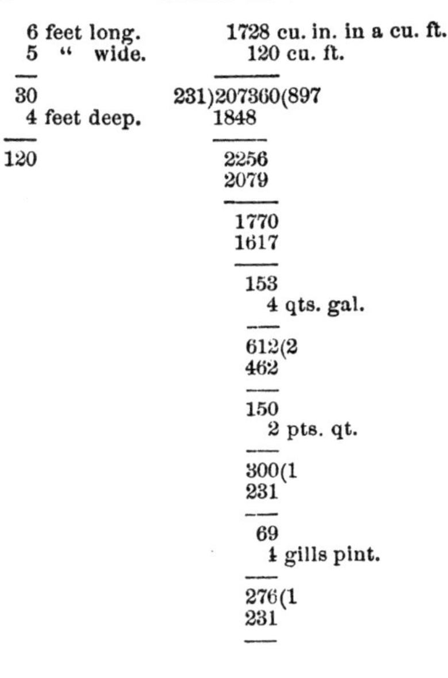

```
6 feet long.            1728 cu. in. in a cu. ft.
5  "  wide.              120 cu. ft.
--                      -------
30                  231)207360(897
 4 feet deep.           1848
--                      ----
120                      2256
                         2079
                         ----
                          1770
                          1617
                          ----
                           153
                             4 qts. gal.
                           ---
                           612(2
                           462
                           ---
                           150
                             2 pts. qt.
                           ---
                           300(1
                           231
                           ---
                            69
                             4 gills pint.
                           ---
                           276(1
                           231
                           ---
```

VULGAR FRACTIONS.

A fraction is a part of any thing and is represented by two numbers, one above the other, and the other below it; thus, $\frac{1}{2}$ $\frac{2}{3}$ $\frac{3}{4}$ read one-half, two-thirds and three-fourths.

The figure above the line is called the numerator; the figure below the line is called the denominator.

The fraction 4-5 read four-fifths, the four is the numerator and the five is the denominator.

The denominator makes the number of equal parts into which the whole number is divided; the numerator shows the number of those intended to be expressed by the fraction thus: if we say that if we have 2-3 of an apple, we mean that the apple was divided into three parts and that we have two of those parts.

A proper fraction is that which its numerator is less than its denominator, as $\frac{1}{2}\ \frac{2}{3}\ \frac{4}{7}$

An improper fraction is that which has its numerator greater than its denominator, as $\frac{3}{2}\ \frac{7}{4}\ \frac{8}{5}$

A compound fraction is a fraction of a fraction, as is expressed by 2 or more fractions, as $\frac{2}{3}$ of or $\frac{1}{3}$ of $\frac{2}{5}$ of $\frac{4}{9}$

A mixed number is a whole number with a fraction annexed, as 2 1-2, 4 2-3, 16 4-5.

Any whole number may be made a fraction of by writing a 1 under it.

A complex fraction is one which has a fraction in its numerator or denominator or both, as

$$\frac{2\ 1\text{-}2}{5}\qquad\frac{4}{5\ 1\text{-}3}\qquad\frac{2\ 1\text{-}3}{8\ 3\text{-}4}$$

ADDITION OF FRACTIONS.

RULE.—Multiply each numerator by all the denominators, except its own, for a numerator, and all the denominators together for a common denominator.

Ex. 1. What is the sum of 3-5 and 5-6?

```
 3 5          3×6=18
 ---          ------
 5 6          5×5=25
                  --
                  43 new numerator
                  --
              5×6 =30 new denominator.
```

Ex. 2. What is the sum of 1-2, 2-5, 3-4 and 2-3?

```
1×5×4×3= 60 the numerator for 1 2.
2×2×4×3= 48  "      "       " 2-5.
3×5×2×3= 90  "      "       " 3-4.
2×4×5×2= 80  "      "       " 2-3.
-----------------
         278
         ---        120)278
2×5×4×3=120         -------
                      2 19-60
```

Ex. 3. What is the sum of 1-20 and 1-35?

```
    1×35=  35
    1×20=  20
    ---------
         5)55 (11 red'd to its lowest terms.
         ---- ---
  20×35= 700  140
```

Note.—The numerator cannot be divided by the denominator. The two numbers are reduced to its lowest terms, that is 5 will divide them without a remainder.

WHOLE NUMBER AND MIXED MUMBER.

RULE.—Reduce them to improper fractions and then proceed as before.

Ex. 4. What is the sum of 12 and 4 1-2?

```
4 1-2                                  9×12
2                                      ----
---                                    2   1
9 red'd to an improper fraction.
---                                  9×1= 9
2                                   12×2=24
          2)33                      -------
          ----                           33
          16 1-2 ans.                    --
                                          2
```

TWO MIXED NUMBERS.

Ex. 5. What is the sum of 4 1-2 and 5 1-2 ?

$$4\tfrac{1}{2} = \frac{9}{2} \qquad 5\tfrac{1}{2} = \frac{11}{2} \qquad \frac{\begin{matrix}9\times2=18\\ 11\times2=22\end{matrix}}{\begin{matrix}40\\ \hline 2\times2=\ 4\end{matrix}} \qquad \frac{4)40}{10} \text{ ans.}$$

COMPOUND FRACTIONS.

Ex. 6. What is the sum of 2-3 of 1-8, 3-5 of 6-3 and 8-7 ?

$$\frac{2}{3} \text{ of } \frac{1}{8}=\frac{2}{24} \qquad \frac{3}{5} \text{ of } \frac{6}{3}=\frac{18}{15} \qquad \frac{2}{24}\ \frac{18}{15}\ \frac{8}{7}$$

```
  2×15× 7= 210
 18×24× 7=3024
  8×15×24=2880          2520)6114
 -------------          ---------
          6114           2 179-420 ans.
 -------------
 24×15× 7=2520
```

Ex. 7. How many feet are there in 6 sides of leather measured as follows: 8 1-4 ft., 6 1-2 ft., 7 3-4 ft., 8 3-4 ft., 9 1-2 ft. and 14 3-4 ft ?

```
 8 feet 1= 8 1-4 feet.
 6  "   2= 6 1-2  "
 7  "   3= 7 3-4  "
 8  "   3= 8 3-4  "
 9  "   2= 9 1-2  "
14  "   3=14 3-4  "
--------------------
55  "   2=55 1-2  "
```

In the 1st column there are 14 quarters=3 1-2 feet, which we carry the 3 feet to the column of feet, making 55 1-2 feet, ans.

TO FIND THE GREATEST COMMON DIVISOR.

RULE.—Divide the greatest number by the less; then divide the divisor by the remainder, and so on, dividing always the last divisor by the last remainder until nothing remains.

A common divisor of two or more numbers is a number which will divide each of them without a remainder.

Thus, 3 is a common divisor of 9, 15 and 18.

The greatest common divisor of two or more numbers, which divide each of them without a remainder; thus, 24, 36, 48, can be divided by 2, 3 or 4, but their greatest common divisor is 12.

Ex. 1. What is the greatest com. div. of $\frac{3132}{6912}$?

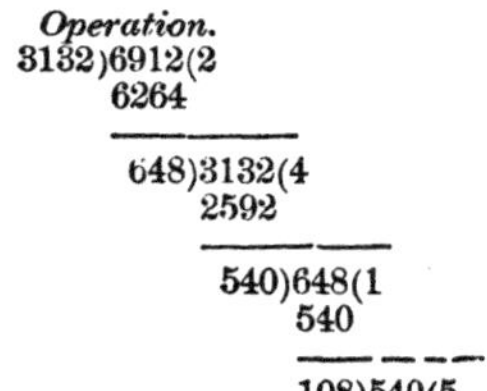

Operation.

```
3132)6912(2
     6264
     ----
      648)3132(4
          2592
          ----
           540)648(1
               540
               ---
               108)540(5
```

Ans., 108 is a divisor common to both terms.

SUBTRACTION OF FRACTIONS.

RULE.—Subtract their numerators, where they have a common denominator, otherwise they must first be reduced to a common denominator.

Ex. 1. From 3-5 take 1-4.

$$3 \dagger \times 4 = 12$$
$$1 \times 5 = 5$$

$$\frac{3}{5} {}^{*}- \frac{1}{4} \qquad \frac{7}{20} \text{ ans.}$$

Ex. 2. From 12-15 take 8-5.

$$\frac{12}{15} - \frac{8}{15} = \frac{5}{15} \text{ ans.}$$

Ex. 3. From 18-27 take 3-9.

$$18 \times 9 = 162$$
$$3 \times 27 = 81$$
$$\frac{81) \quad 81(1\text{-}3}{27 \times 9 = 243} \text{ ans.}$$

WHOLE NUMBER AND A FRACTION.

Ex. 4. From 20 take 3-5.

$$\frac{20}{1} - \frac{3}{5} \qquad \begin{array}{r} 20 \times 5 = 100 \\ 3 \times 1 = 3 \\ \hline 97 \\ \hline 1 \times 5 = 5 \end{array} \qquad \begin{array}{l} 5)97 \\ \hline 19\ 2\text{-}5 \text{ ans.} \end{array}$$

TWO MIXED NUMBERS.

RULE.—When one mixed number is to be subtracted from another, we may reduce both numbers to an improper fraction, and then to a common denominator.

*—Sign of subtraction. †x sign of multiplication.

9

Ex. 5. From 9 1-4 take 7 3-4.

```
 9 1-4      7 3-4
 4          4                                  37—31
37          31 red'd to an improper frac.      4   4
 4          4
              37×4=148
              31×4=124
                    24        16)24
                    16          1 1-2 ans.
```

WHOLE NUMBER AND A MIXED NUMBER.

Ex. 6. From 5 take 4 1-2.

```
 4 1-2                        5×2=10
 2            5—9             9×1= 9
 9            1 2                  1  ans.
 2                            1×2= 2
```

MULTIPLICATION OF FRACTIONS.

RULE.—Prepare the fractions as previously required, multiply the numerators together for a new numerator, and denominators together for a new denominator.

MULTIPLY A FRACTION BY A FRACTION.

Ex. 1. Multiply 4-5 by 7-8.

```
4*×7=4)28( 7 ans.
5 ×8   40 10
```

MULTIPLY A FRACTION BY A WHOLE NUMBER.

RULE.—Multiply the numerators of the fraction by the whole number and write the product over the denominator.

*Sign of multiplication.

Ex. 2. Multiply 21-30 by 15.

$\frac{21}{30} \times \frac{15}{1} = \frac{315}{30}$ 30)315 10 1-2 ans.

WHOLE NUMBER AND A MIXED NUMBER.

RULE.—Multiply the whole number by the denominator of the fraction, and to the product add the numerator, then set that sum above the denomator.

Ex. 3. Multiply 50 by 4 1-2.

4 1-2
2
9
2

$\frac{50}{1} \times \frac{9}{2} = \frac{450}{2}$ 2)450 225 ans.

TWO MIX NUMBERS.

Ex. 4. Multiply 125 1-2 by 7 3-4.

125 1-2 7 3-4
2 4
251 31
2 4

$\frac{251}{2} \times \frac{31}{4} = \frac{7781}{8}$ 8)7781 972 5-8 ans.

Ex. 5. If one yard broad cloth cost $4 50, what will 1-8 of a yard cost? What will 1-16 cost?

Ans. to the 1st, 56 1-4 cts.
Ans. to the 2d, 28 1-8 cts.

$\frac{1}{8} \times \frac{450}{1} = \frac{8)450}{8}$ = 56 1-4 cts.

$\frac{1}{16} \times 450 =$ 16)450 28 1-8 cts.

DIVISION OF FRACTIONS.

RULE.—Prepare the fractions as in multiplication; then invert the divisor and proceed as in multiplication.

DIVIDE A FRACTION BY A FRACTION.

Ex. 1. Divide 42-54 by 24-35.

$\frac{42}{54} \div \frac{24}{35}$ inverted thus, $\frac{42}{54} \times \frac{35}{24}$ $\frac{1470}{1296}$

1295)1470
 129-216 ans.

Ex. 2. Divide 4-7 by 3-5.

3-5 inverted thus, $\frac{4}{7} \times \frac{5}{3} = \frac{20}{21}$ ans.

COMPOUND FRACTIONS.

When compound fractions occur in the divisor or dividend, they must be reduced to simple ones, and mixed numbers must be reduced to an improper fraction.

Ex. 3. Divide 3-4 of 2-3 by 2 1-3.

2 1-3
$\frac{3}{}$
$\frac{7}{3}$ improper frac.

$\frac{3}{4}$ of $\frac{2}{3} = \frac{6}{12}$ reduced to si'le ones.

$\frac{6}{12} \times \frac{7}{3}$ inverted thus $\frac{6}{12} \times \frac{3}{7} = 6)\frac{18}{84}(3\text{-}14$ ans.

WHOLE NUMBER AND A MIXED NUMBER.

Ex. 4. Divide $28 by 3 1-2.

1st method.

```
3 1-2        2800
2               2
—            ————
7           7)5600
—            ————
2           $8.00 ans.
```

2d method.

```
28×7 inverted thus, 28×2=56     7)56
———————————————————————————     ————
 1  2                1  7  7      8 ans.
```

TWO MIXED NUMBERS.

Ex. 5. Divide 8 2-3 by 3 1-2.

```
                                     8 2-3    3 1-2
                                     3        2
                                     ——————————————
26  7 inverted thus, 26×2=52         26       7
————————————————————————————         ——————————————
3   2               3  7  21         3        2

                      21)52
                      —————
                      2 10-21 ans.
```

The numerator represents the dividend and the denominator the divisor.

DIVIDE A FRACTION BY A WHOLE NUMBER.

RULE.—There are two ways; 1st, divide the numerator by the whole number (if it will contain it without a remainder), and under the quotient write the denominator. Otherwise multiply the denominator by the whole number, over the product write the numerator.

Ex. 6. Divide 21-25 by 7.

```
7)21
————
3-25 ans.
```

DIVIDE A WHOLE NUMBER BY A FRACTION.

RULE.—Multiply the whole number by the denominator, and divide the product by the numerator.

Ex. 7. Divide 75 by 5-9.

```
   75
    9
   --
5)675
  ----
  135 ans.
```

Ex. 7. Divide 120 by 10 3-4

```
10 3-4          120
 4                4
                 --
------        43)480
43-4            -------
                11 7-43 ans.
```

ADDITION OF DECIMAL FRACTIONS.

RULE.—Write the numbers so that the same order may stand under each other; placing tenths under tenths and hundredths, etc.

Begin at the right hand or lowest order, proceed in all respects as in adding whole numbers.

The decimal point in the answer will always fall directly under the decimal points in the given number.

Ex. 1. What is the sum of 3,5; 25,467; 125,4 and 2,466?

```
  3,5
 25,467
125,4
  2,466
-------
156,833 ans.
```

Write the units under units, tenths under tenths, hundredths under hundredths, etc.; then beginning at the right hand or lower order, proceed thus: 6-thousandths and 7-thousandths are 13-thousandths.

Write the 3 under the column added and carry the 1 to the next column, as in addition of whole numbers; one to carry to 6-hundreths makes 7-hundreths, and 6 are 13-hundreths, set the 3 under the column and carry the 1 as before; 1 to carry to 4-tenths makes 5 and 4 are 9-tenths, 4 are 13-tenths and 5 are 18-tenths or 1 and 8-tenths.

Set the 8 under the column and carry the 1 to the next column. Finally, place the decimal point in the amount directly under that number added.

Decimal fractions are commonly expressed by writing the numerator only, with a point (.) before it, called the decimal point, thus:

9-10	is written thus,		.9
99-100	"	"	.99
999-1000	"	"	.999

The denominator to a decimal fraction, although not expressed, is always understood, and is 1 with as many ciphers annexed as there are places at the right hand of the point, thus: .99 is a decimal of two places, consequently 1, with 2 ciphers annexed, (100), is its proper denominator.

SUBTRACTION OF DECIMAL FRACTIONS.

RULE.—Write the lower number under the greater with units under units, tenths under tenths and so on. Subtract as in whole numbers.

Ex. 1. From 26,467 subtract 14,18

```
26,467
14,18
------
12,287
```

Having written the less number under the greater, units under units, etc.; thus, 0-thousandths from 7-thousandths, leaves 7-thousandths, with the 7 in the thousandths place; as the next figure in the lower one is larger than the one above it, we borrow 10. Now 8 from 16 leaves 8; set the 8 under the column and carry 1 to the next figure. Proceed in the same manner with the other figures in the lower number; place the decimal point in the remainder under that in the given numbers.

Ex. 2. From 16 take 1.5

```
16.0
 1.5
----
14.5 ans.
```

Ex. 3. From 1 take .125.

```
1
0125
----
.875 ans.
```

Ex. 4. From .56078 take .40003.

```
.56078
.40003
------
.16075 ans.
```

MULTIPLICATION OF DECIMAL FRACTIONS.

RULE.—We multiply as in whole numbers and point off as many decimals in the product as there are decimal figures in both factors.

Ciphers on the left of decimals do not effect their value. The 0 may be omitted.

Ex. 1. Multiply .48 by .5.

```
 .48
 .5
----
.240
```

Ex. 2. Multiply 8.46 by .25

```
  8.46
   .25
------
2.1150
```

Ex. 3. Multiply .045 by .03.

```
  .045
   .03
------
.00135
```

In the example No. 3, there are 5 decimal places in the factors and only three figures in the product, therefore two ciphers are placed at the left of the product to make the number of decimal places equal to those in the factors.

DIVISION OF DECIMAL FRACTIONS.

RULE.—Divide as in whole numbers. Point off as many decimal places in the quotient as the dividend has more than the divisor.

When the decimal places in the divisor exceed those in the dividend, make them equal by annexing ciphers to the right of the divldend, and the quotient will be a whole number.

When there is a remainder, the quotient may be carried to any degree of exactness by annexing ciphers to the remainder.

Ex. 1. Divide 4.7614 by 3.8.

3.8)4.7614
1.253

Ex. 2. Divide .7644 by .42.

42).7644
.0182

In the first example the decimals in the dividend exceed those in the divisor by 3; three figures are, therefore, marked off in the quotient.

And in the second example the decimals in the dividend exceed those in the divisor by 4; one cipher is therefore prefixed in the quotient to make four decimal places.

Ex. 3. Divide .289 by 2.4.

2.4).289
.12+ans.

The decimal places exceed those in the divisor by 2; we point off two decimal places in the quotient.

When there is a remainder the sign (+) of addition should be annexed to the quotient to show that it is not complete.

Ex. 4. Divide .063 by 9.

9).063
.007 ans.

In this example the dividend has 3 more places of decimals than the divisor; the quotient must have 3 places of decimals. We must therefore prefix 2 ciphers to the quotient.

TO READ A DECIMAL FRACTION.

Beginning at the left hand, read the figures as if they were whole numbers, and the last one, add the name of its order.

Thus,	.7	is read	7-tenths.
	.36	"	36-hundreths.
	.475	"	475-thousandths.
	.6342	"	6342-ten thousandths.
	.57834	"	57834-hundred thousandths.
	.284648	"	284648-millionths.
	.8913629	"	8913629-ten millionths.

All the figures which come before the point are whole yards, pounds, acres, etc., as the case may be. All which come after the point are fractions of the same. Thus, 1.7 pounds stands for 1 pound and 7-tenths of a pound; 12.34 yards stand for 12 yards, 3-tenths of a yard and 4-hundredths of a yard, or 12 yards and 34-hundreths of a yard.

DUODECIMALS OR CROSS MULTIPLICATION.

This rule is particularly useful to glaziers, masons, marble and slate dealers.

One foot is divided into 12 equal parts, called inches or primes, and marked ( ).

Each is again divided into 12 equal parts called seconds, and marked (″).

Each second is again divided into 12 equal parts called thirds, and marked (‴).

Each third is again divided into 12 equal parts called fourths, and marked (⁗)

— —

DUODECIMALS.

Under the multlplicand write the corresponding denomination of the multiplier. Multiply each term of the multiplicand by each term of the multiplier in succession, beginning at the lowest denomination, observing to carry a unit for every twelve from each denomination to the next higher.

The sum of these partial products will be the answer required.

By this rule, also may be calculated the superficial

and solid contents of bodies, having the measures of their different sides.

The more easily to comprehend the rule—

Feet	multiplied	by feet	give	feet.
"	"	inches	"	inches.
"	"	seconds	"	seconds.
Inches	"	inches	"	"
"	"	seconds	"	thirds.
Seconds	"	"	"	fourths.

Ex. 1. Multiply 8 feet 5′ by 7 feet 3′, or how many superficial feet. Ans. 61 ft. 0′ 3″.

We begin on the right hand and multiply each denomination of the multiplicand, first by the inches of the multiplier and by the feet of the multiplier as follows:

Feet.	Inches.		
8	5′		
7	3′		
2	1′	3″	
58	11′		
61	0′	3″	super'l ft. ans.

3′×5′=15″=1′ 3″ we write down the 3″ and reserve the 1′ for the next product.

Again, 8 ft.×3′=adding in the 1′ which was reserved from the last product, we have 25′=2 ft. 1′, which we write down entire.

Again, we have 7 ft.×5′=35′=2 ft. 11′; we write down the 11 under the primes of the former line, and reserved the 2 ft. for the next product; 8 ft.×7 ft.=56 ft. to which, adding the 2 ft. reserved from the last product, we have 58 feet, which we place under the feet of the former line; taking the sum we have 61 ft. 0′ 3″ for the answer.

Ex. 2. Multiply 8 feet by 3 feet 4 inches, by 4 feet 2 inches.

Feet.	Inches.		
8	0′		
3	4′		
2	8′	0″	
24	0′		
26	8′	0″	
4	2′		
4	5′	4″	0‴
106	8′	0″	
111	1′	4″	0‴

Multiply as follows:—4×0=0″; we write down the 0″

Again, 8 ft.×4′=32′=2 ft. 8′, which we write down entire.

Again, we have 3 ft.×0′=0′; we write down the 0′ under the inches of the former line; 8 ft.×3 ft.=24 ft. we place under the feet of the former line.

Taking the sum we have 26 ft., 8′ 0″; again, 2′×0″ =0, we write down the 0‴; 2×8′=16″=1′ 4″, we write down the 4″ and reserve the 1′ for the next product.

Again, 26 ft.×2′=52′, adding in the 1′ which was reserved from the last product, we have 53′=4 ft. 5′, which we write down entire.

Again, $4 \text{ ft.} \times 0'' = 0''$, we write down under the seconds of the former line.

Again, we have $4 \text{ ft.} \times 8' = 32' = 2 \text{ ft. } 8'$; we write down the $8'$ and reserve the 2 ft. for the next product.

Again, $26 \text{ ft.} \times 4 \text{ ft.} = 104$ ft., adding in the 2 ft. reserved from the last product, we have 106 ft. which we place under the feet of the last former line. Taking the sum, we have 111 ft. $1'\ 0''$ for the answer.

Ex. 3. Multiply 7 feet, 3 inches, 2 seconds by 1 foot, 7 inches and 3 seconds.

Feet.	inch.	seconds.		
7	$3'$	$2''$		
1	$7'$	$3''$		
	$1'$	$9''$	$9'''$	$6''''$
4	$2'$	$10''$	$2'''$	
7	$3'$	$2''$		
11	$7'$	$9''$	$11'''$	$6''''$

Multiply as follows: $2'' \times 3'' = 6''''$; we write down $6''''$. Again, $3'' \times 3' = 9'''$; we write down $9'''$. Again, $7 \text{ ft.} \times 3'' = 21'' = 1'\ 9''$; we write down entire. Again, $7' \times 2'' = 14''' = 1''\ 2'''$; we write down the $2'''$ and reserve the $1''$ for the next product.

Again, $3' \times 7' = 21''$, adding in the $1''$ reserved from the last product, we have $22'' = 1'\ 10''$; we write down $10''$ and reserve $1'$ for the next product.

Again, $7 \text{ ft.} \times 7' = 49'$, adding in the $1'$ reserved from the last product, we have $50'' = 4 \text{ ft. } 2'$, which we write down entire.

Again, $1 \text{ ft.} \times 2'' = 2''$, we unite $2''$ under the last former line.

Again, 1 ft.×3′=3′, we write down under the primes of the former line.

Again, 7 ft.×1 ft.=7 ft., we write down under the feet of the last former line.

Taking the sum, we have 11 ft. 7′, 9″, 11‴, 6⁗, the answer.

Ex. 4. Multiply 2 feet by 11 inches by 6 inches.

```
  2   0′
     11
-------------
  1  10    0″
      6′
-------------------
     11′   0″   0‴
```

Multiply as follows: 11′×0′=0″; we write down 0″.

Again, 2 ft.×11′=22′=1 ft. 10′; we write down entire.

Again, 6′×0″=0‴; we write down 0‴ under the last former line.

Again, 10′×6′=60″=5′ 0″; we write down 0″ and reserve the 5′ for the next prodoct.

Again, 1 ft.×6′=6′, adding in the 5′ reserved from the last product, we have 11 inches the answer.

Ex. 5. Multiply 2 ft. 2 inches by 1 foot by 6 inches.

```
  2   2′
  1   0′
-----------
  2   2′
      6′
--------------
  1   1′   0″
```

Multiply as follows: 1 ft.$\times$2′=2′; we write down 2′.

Again, 1 ft.$\times$2 ft=2 ft; we write down.

Again, 2′$\times$6′=12″=1′ 0″; we write down the 0″.

Again, 2 ft.$\times$6′=12′, adding in 1′ which was reserved from the last product, we have 13′=1 ft. 1 inch, the answer.

Ex. 6. How many cubic feet are there in 3 ps. slate 2 feet by 10 inches by 6 inches?

2	2′			
	10′			
1	9′	8″		
	6′			
	10′	10″	0‴	
			3 pieces	
2	8′	6″	0‴	ans.

Multiply as follows: 2′$\times$10′=20″=1′ 8″; we write down 8″ and reserve the 1′ for the next product.

Again, 2 ft.$\times$10′=20′, adding in the 1 reserved from the last product, we have 21′=1 ft. 9′, which we write down entire.

Again, 6′$\times$8″=48‴=4″ 0‴; we write down 0‴ and reserve 4″ for the next product.

Again, 6′$\times$9′=54″, adding in 4″ reserved from the last product, we have 58″=4′ 10″; we write down the 10″ and reserve the 4′ for the next product.

Again, 1 ft.$\times$6′=6′, adding in the 4′ reserved from the last product, we have 10′ which we write down.

Again, 3 ps.$\times$0‴=0‴, we write down.

Again, 3 ps.×10″=30″=2′ 6″; we write down the 6″ and reserve the 2′ for the next product.

Again, 3 ps.×10′=30′, adding in 2′ reserved from the last product, we have 32′=2 ft. 8′, which we write down entire for the answer.

TRIANGLES TO FIND THE THIRD SIDE.

RULE.—First add together the square of the base and of the perpendicular, and the square root of the sum is the hypotenuse or longest side.

RULE 2d—Add together the hypotenuse and any one side, multiply the sum by their difference, and the square root of the product equals the other side.

Ex. 1. A wall, 36 feet high, and a ditch before it is 27 feet wide; what is the length of a ladder that will reach to the top of the wall from the opposite side of the ditch.

```
  36                     2025(45 feet ans.
  36          27         16
 ----         27         ----
 1296        ---     85) 425
  729        729         425
 ----                    ----
 2025
```

Note.—A figure of the sides like that formed by the wall, the ditch and the ladder, is called a right angle triangle, of which the square of the hypotenuse, or slanting side, (the ladder), is equal to the sum of the square of the other two sides that the height of the wall and the width of the ditch.

Ex. 2. A line of 36 yards will exactly reach from the top of a fort to the opposite bank of a river, known to be 24 yards broad; the height of the wall is required.

```
                                720(26.83
                                4
                                ---
36                            46)320
24            36                 276
--            24                 ----
60            --              528)4400
12            12                  4224
---                               ------
720                          5363) 17600
                                   16089
                                   -----
```

Ans. 26 yards and 83 hundredths of a yard.

CIRCLES.

Rule.—To find the length of the diameter, multiply the length of the diameter by 3 1-7, the product will be the circumference.

To find the diameter, divide the circumference by 3 1-7, and the quotient will be the diameter.

To find the area or surface of a circle, multiply the square of the diameter by .7854, the product will be the area.

Ex. 1. If the diameter is 231 feet, what is the circumference?

```
7)231
   3 1-7
 -----
 693
  33
 ----
 726 feet ans.
```

Ex. 2. If the circumference is 726 feet, what is the diameter?

```
3 1-7          726
7                7
----           ----
22            )5082
               ----
               231 feet ans.
```

Ex. 3. What is the area of a circle whose diameter is 6 inches? Ans. 28 ft.

```
 6      .7854
 6         36
--     -------
36    28.2744
```

SQUARE ROOT.

Rule with Example.—Divide the given number into periods of two figures each by placing a point over the unit figure, and every alternate figure toward the left.

1st, Find the square root of 3 of the first period, 10, and place it in the quotient.

Subtract the square of it, 9, from the first period, and to the remainder annex the next period, 69, for a dividend.

Double the root already found, 3, for a divisor, and supposing the unit figure 9, omitted, find how often 6 is contained in the dividend. It is containtd 2 times; place the 2 in the quotient and the divisor; multiply by 2 the divisor 62, ar d subtract the product 124 from the dividend.

2d, Double the figures already found in the root for a new divisor, (or bring down your last divisor for a new one, doubling the right hand figure of it), and from these, find the next figure, in the root, as last directed, and continue operation in the same manner till you have brought down all the periods.

If the divisor is not contained in the dividend, place a cipher in the root; also, on the right of the divisor, and bring down the next period.

If there is a remainder after the periods are brought down, periods of ciphers may be annexed, and the figures of the root thus obtained will be decimals.

Ex. 1. What is the square root of

```
      106929(327
      9
      ----                 proof.
  62) 169                   327
      124                   327
      ----                -------
 647) 4529                106929
      4529
      ----
```

Ex. 2. What is the square root of 164?

```
     164(12.8 ans.
     1
     -
  22)64
     44
     --
 248)2000
     1984
     ----
```

Ex. 3. What is the square root of 10308921894001? Ans. 3210751.

```
                    10308921894001(3210751
                     9
                     --
1st divisor 62)130
               124
               ---
2d     "    641)  689
                  641
                  -------
3d     "    64207) 482198
                   449449
                   -------
4th    "    642145) 3274940
                    3210725
                    --------
5th    "    6421501)  6421501
                      6421501
                      -------
```

Ex. 4. What is the square root of 4-25? Ans. 2-5.

```
  4(2            25(5
  4             5)25
  -               --
2)0
```

Ex. 5. What is the square root of 20 1-4?

```
          20 1-4        2025(4.5=4 1-2 ans.
4)81       4          4)16
-----      --           ---
 20.25     81        85) 425
           --            425
            4
```

CUBE ROOT.

Rule with Example.—Divide the given numbers into periods of three figures, beginning at the place of units. Place the cube root of the first period 2, in the quotient, and subtract its cube, 8, from the first period, and bring down the next period for a dividend, which is 4812.

To find a divisor, multiply the square of the figure placed in the quotient by 300=1200; find how often this is contained in the dividend, viz: 3 times; place the 3 in the quotient for the second figure of the root.

Multiply the part of the root formerly found, viz: 2, by the last figure placed in the root, viz: 3, and the product by 30=180; add this and the square of the last figure placed in the root to the divisor, viz: 1200.

Multiply the sum of these, 1389, by the last figure placed in the root 3, and subtract the product, 4167, from the dividend. 4812; bring down another perjod for a new dividend and proceed in the same manner.

Ex. 1. Find the cube root of

```
                          1̇2812̇904̇(234
          2×2=4×300=1200   8
          2×3=6× 30= 180  ----          proof.
          3×3=         9   4812            234
          --------------                   234
               1389×3=     4167
                          ------         -----
23×23=529×300=158700         645904      54756
23× 4= 92× 30=  2760                       234
 4× 4=      16=   16                  --------
-------------------
        161476×4=645904                12812904
```

Proof.—Multiply the root into itself twice, and if the last product is equal to the given number, the work is right.

When there is a remainder, periods of ciphers may be added, and the figures of the root thus obtained, will be decimals.

If the right hand period of decimals is deficient, this deficiency must be supplied by ciphers.

When there are decimals in the given example, find the root as in whole numbers; then point off as many decimal figures in the answer as there are periods of decimals in the given number.

If the divisor is not contained in the dividend, place a cipher in the root; also, two ciphers on the right of the divisor, and bring down the next period.

In finding the cube root of a common fraction, first reduce the fraction to its lowest terms, then extract the root of its numerator and denominator.

When either the numerator or denominator is not a perfect cube, the fraction should be reduced to a decimal, and the root of the decimal be found as above.

A mixed number should be reduced to an improper fraction.

Ex 2. What is the cube root of 27-64? Ans. 3-4.

27(3 numerator. 64(4 denominator.
27 64
— —

Ex. 3. What is the cube root of 13 2-3?

13 2-3 3)41000000000
3 ———————
— 13666666666
41
—
3 13666666666(2.3908 ans
8
———
2×2=4×300=1200 5666
2×3=6× 30= 180
3×3= 9
———————
1389×3=4167
———
1499666

```
23×23=529×300=158700
23× 9=207× 30=   6210
 9× 9=             81
                164991×9=1484919

                         147476660
239×239=57121×300=17136300
239×  0=  239× 30=    7170
  8×8                   64
                17143534×8=137148272
```

Ex. 4. The dimensions of a round bushel measure are 18 1-2 inches wide and 8 inches deep; what will be the dimensions of a similar measure that will hold 8 bushels?

Ans. 37 inches wide, 16 inches deep.

```
    18.5                             8
    18.5                             8
  342.25                            64
    18.5                             8
 6331625            50653(37       512
       8 bu.        27               8 bu.
50653000            23653         4096

  3×3= 9×300=2700
  3×7=21× 30= 630
  7×7=         49
           3379×7=23653

  1×1=1×300=300
  1×6=6× 30=180            4096(16
  6×6=       36            1
           516×6= 3096     3096
                           3096
```

MEASUREMENT OF SURFACES.

How many acres are there in a farm 416 rods long and 170 rods wide?

```
                     416
                      170
                     -----
160 rods an acre)70720
                     -----
                      442 acres ans.
```

Ex. 2. How many square feet are there in a lot which is laid out in a right angle triangle, the base measuring 49 feet, and the perpendicular 30 feet?

```
      49
       30
     -----
  2)1470
     -----
      735 ans
```

Rule.—Multiply the base by the perpendicular, and divide the product by 2; or multiply the base by one-half of the perpendicular.

Ex. 3. There is a lot 75 feet wide and 8 rods long, how many square rods does it contain?

Ans. 36 36-100 rods.

```
272 1-4 sq. ft. 1 sq. rod.        16 1-2 ft. rod.
  4                                  8 rods long.
-----                              ----
1089 quarters.                     132
                                    75 feet wide.
                                   ----
                                   9900
                                      4
                                   ----
                            1089)39600 quarters.
                                   ------
                                   36.36
```

Ex. 4. There is a lot of land 25 feet long and 75 feet wide, and was sold for $6575; what did it cost per square foot?

```
   75
   25                1875)6575.00
 ----                       -------
 1875 sq. feet.            $3.50.6 ans.
```

Point off in the quotient the same number of decimals as Ex. 3, p. 27.

Ex. 5. How many acres in a tract of land 20 miles broad and 250 miles long?

```
   250
    20
  ----
  5000
   640 acres in a square mile.
---------
3,200,000
```

Ans., three million two hundred thousand acres.

Ex. 6. What is the area of a garden which is 18 rods long and 15 rods wide?

```
 18
 15
 --
270 rods ans.
```

Ex. 7. How many feet in 200 yards?

```
200
  3 feet a yard.
 --
600 feet the ans.
```

Ex. 8. How many acres in 2118165 1-2 yards?

```
2118165 1-2
      2
------- inverted.
4236331 × 4   = 16945324
-------   ---   --------
   2      121      242
```

```
          242)16945324            30 1-4 sq. yds a sq. rod.
              --------             4
16 rods an acre.) 70022           --
                  -------         121
                   437.63         ---
                                   4
```

Four hundred, thirty-seven acres and sixty-three hundredths of an acre.

DENOMINATE NUMBERS.

STERLING MONEY.

4 farthings make 1 penny,	marked	d.
12 pence a shilling,	"	s.
20 shilling make a pound,	"	£
21 " " guinea,	"	g.

TROY WEIGHT.

Troy weight is used for weighing gold, silver, jewels, etc., and for ingredients used in philosophical experiments.

	marked
24 grains (g) make a pennyweight,	pwt.
20 pennyweights make 1 ounce,	oz.
12 ounces " 1 pound,	lb.

AVOIRDUPOIS WEIGHT.

By this weight, coarse bulky goods are weighed and, all the common necessaries of life.

	marked
16 drams (dr.) make 1 ounce,	oz.
16 ounces make 1 pound,	lb.
25 pounds make 1 quarter,	qr.
4 quarters (or 100 lbs.) hundred weight,	cwt.
20 hundred weight 1 ton,	T.

Apothecaries and chemists use this weight in mixing medicines, but drugs are bought and sold by avoirdupois weight.

	marked
20 grains (g) make 1 suruple,	sc.
3 scruple make 1 dram,	dr.
8 drams make 1 ounce,	oz.
12 ounces make 1 pound,	lb.

LONG MEASURE.

	marked
12 inches (in.) make 1 foot,	ft.
3 feet make 1 yard,	yd.
5 1-2 yards, 16 1-2 ft. 1 rod, perch or pole,	r. or p.
40 rods make 1 furlong,	fur.
8 furlongs, or 320 rods make a mile,	mi.
3 miles make on league,	lea.
60 geographical miles or 69 1-2 statute miles make 1 degree,	deg.
360 degrees make a great circle, or circumference of the earth.	

CLOTH MEASURE.

Cloth measure is used in measuring cloth, carpeting, ribbons, etc.

	marked
2 1-4 inches (in.) make 1 nail,	na.
4 nails, or gin, 1 quarter yard,	qr.
4 quarters make 1 yard,	yd.
3 qrs, 3-4 of a yard, 1 Ell Flemish,	Fl.
5 qrs., or 1 1-4 makes 1 Ell English,	E'e.
6 qrs., or 1 1-2 yards make 1 Ell French,	F.

SQUARE MEASURE.

Square measure is used for measuring surfaces, such as land, paving, flooring, plastering, boards, etc.

	marked.
144 square inches (sq. in.) make 1 sq. ft.	sq. ft.
9 square feet make 1 square yard.	sq. yd.
30 1-4 square yards, or 272 1-4 feet, 1 sq. rod, perch or pole,	sq. r.
40 square rodc 1 square rood,	R.
4 roods, or 160 square rods, 1 acre,	A.
640 acres 1 square mile,	M.

CUBIC MEASURE.

Cubic measure is used in measuring solid bodies or spaces; that is, things having length, breadth and height, or thickness, such as earth, stone, timber, boxes of goods, the capacity of rooms, etc.

	marked
1728 cubic inches (cu. in.) make a cubic foot,	cu. ft.
27 cubic feet make a cubic yard,	cu. yd.
40 feet of round timber, or 50 feet of hewn make 1 ton or load,	T.
42 cubic feet 1 ton of shipping,	T.
16 cubic feet 1 foot of wood or cord foot.	ft.
8 cord feet, or 128 cubic feet, 1 cord,	c.

BEER MEASURE

Is used in measuring beer, ale and milk.	marked
2 pints make 1 quart,	qt.
4 quarts make 1 gallon,	gal.
36 gallons make 1 barrel,	bbl.
1 1-2 barrels, or 54 gallons, 1 hogshead,	hhd.

LIQUID MEASURE.

The standard gallon of the United States measures 231 cubic inches, and contains 833888 pounds avoirdupois of distilled water. The British imperial gallon contains 10 avoirdupois of distilled water and measures 277 1-4 cubic inches.

	marked
4 gills (gi.) make 1 pint,	pt.
2 pints make 1 quart,	qt.
4 quarts make 1 gallon,	gal.
31 1-2 gallons make 1 barrel,	bar.
63 gallons make 1 hogshead,	hhd.
2 hogsheads make 1 pipe,	pi.
2 pipes make 1 ton,	ton.
16 large tablespoonfuls are	1-2 pint.
8 " "	1 gill.
4 " "	1-2 gill.
2 gills make	1-2 pint.
A common sized tumbler is	1-2 pint.
" " teacup is	1 gill.
" " wine glass is	1-2 gill.
A large wine glass,	2 ounces.
A tablespoonful makes	1-2 an ounce.

40 drops are equal to one teaspoonful.

4 teaspoonfuls are equal to one tablespoonful.

MISCELLANEOUS TABLE.

12 individual things make a dozen.
12 doz, 144 units, make 1 gross.
12 gross, 1728, make 1 great gross.
20 individual things make 1 score.
56 pounds make 1 ferkin of butter.
100 pounds make 1 quintal of fish.
196 pounds make 1 barrel flour.
200 pounds make 1 barrel pork.
18 pounds make 1 cubit.
80 pounds make 1 bushel salt.
24 sheets of paper make 1 quire.
20 quires make 1 ream.
4 inches 1 hand—used to measure horses.
6 feet a fathom—used to measure depths at sea.
3 miles a league—used reckoning distances at sea.
1 cubic foot marble will weigh 180 pounds.
Marble 2 inches thick 31 1-2 lbs. for every foot.
Marble 1 1-4 in. thick will weigh 20 lbs. for every foot.
Marble 1 1-2 in thick will weigh 24 lbs. for every foot.
Heaped bushels, anthracite coal, weighs 80 lbs.
Marble 1 inch thick will weigh 13 lbs. for every foot.
14 pounds of iron or lead make 1 stone.
21 1-2 pounds of stone make 1 pig.
1760 yards, 5280 ft. or 80 chains, make 1 mile.
5760 grains, Troy, make 1 pound Troy.
7000 " " " 1 pound avoirdupois.
437 1-2 " " " 1 ounce "
27 11-32 " " " 1 dram "
27 cubic feet for a ton of 2000 lbs. hard coal.
Wheat flour, 1 pound is 1 quart.
Indian meal, 1 pound is 2 quarts.
Butter, when soft, 1 pound is 1 quart.

WEIGHTS OF A BUSHEL PRODUCE.

Clover seed,	60	Corn on cob,	70
Beans,	62	Barley,	48
Buckwheat,	48	Flax seed,	45
Indian corn,	56	Oats,	32
Timothy seed,	44	Peas,	60
Potatoes,	60	Sweet Potatoes,	56
Rye,	56	Wheat,	60

NUMERATION TABLE.

Numeration is the art of reading numbers when expressed by figures.

9 Hundreds of trillions,
2 Tens of trillions,
4, Trillions,
6 Hundreds of billions,
7 Tens of billions,
1, Billions,
8 Hundreds of millions,
6 Tens of millions,
2, Millions,
4 Hundreds of thousands,
7 Tens of thousands,
9, Thousands,
5 hundreds,
3 Tens,
8 Units.

The above numbers will read nine hundred and twenty-four trillions, six hundred and seventy-one billions, eight hundred and sixty-two millions, four hundred and seventy-nine thousand, five hundred and thirty-eight.

BANKRUPTCY.

A bankrupt is a person who is insolvent or unable to pay his debts.

Assets is effects of an insolvent person—stock in trade.

Ex. 1. The liabilities of a bankrupt are $63240, and his assets are 12648; what per cent can he pay on the dollar.

```
        12648
          100
        -------
63240)1264800
        -- ---
           20 per cent. ans.
```

Ex. 2. A bankrupt compromises with his creditors at 64 cts on the dollar; how much did B receive on his debt of 2563.50?

```
  2563.50
      .64
 ---------
$1640·64.00
```

Point off from the product the same as Ex. 8, p. 3.

Ex. 3. A man failed in business owes A $156.45, B 256.40 and C $360.40, and his effects are valued at $317; how much will each receive?

```
156.45       773.25 : 156.45 :: 317
256.40                  317
360.40               -------
------     773.25) 49594.65
773.25             ---------
                     64,137
```

773.25 : 256.40 :: 317
317

773.25) 81278.80

105,113

773.25 : 360.40 :: 317
317

773.25)114246.80

147,749

A's 64,138 nearly.
105,113
147,749

$317,000

Ex. 4. How much can a bankrupt pay on the dollar who has $6540 real estate, and owes $56000?

Ans. 11 cts., 6 mills.

6540
100

56000)654000

11.6

Point off in the quotient in the same manner as Ex. 11, in assessment of taxes, p. 96.

GOODS BOUGHT AND SOLD BY THE TON.

Rule.—Multiply the number of pounds by one-half the price* and divide the product by 1000, that will point off 3 figures from the product, and two more for cents; all the figures at the left hand will be dollars, and those at the right hand will be cents.

To find the price per hundred, annex 2 ciphers to the price per ton, divide by two and point off in the quotient in the same manner as above.

*Multiply by one-half the price, because it cost double the price at 1000 pounds, as it would at 2000 pounds.

Ex. 1. What will 960 pounds of hay come to at $15 per ton?

```
     960  pounds.
      750=1-2 the price.
 ──────────
 7.20.000
```

Ex. 2 What will 48040 lbs. egg coal amount to at $15.50 per ton? Ans. $372.31.

```
     48040
      7.75=1-2 the price.
 ────────────
 372.31000
```

Ex. 3. What will 60 pounds of sugar cost at $6.50 per ton? Ans. 19 cents.

```
 3.25=1-2 the priee.
    60
 ───────
 .19.500
```

Ex. 4. What will 1252432 tons of coal amount to at $5 per ton?

```
   1252432
         5
 ─────────
 6,262,160
```

Six millions, two hundred and sixty-two thousand one hundred and sixty dollars the ans.

GOODS BOUGHT AND SOLD BY THE HUNDRED.

RULE.—Multiply the number of pounds by the price, divide the product by 100; that will point off two figures from the product, and then point off two more figures for cents; the figures at the left hand will be dollars, and those at the right will be dollars and cents.

Ex. 1. What will the freight of a box of goods come to weighing 300 lbs., at 30 cents per hundred?

Ans. 90 cents.

```
  300
   30
 ----
,90,00
```

Ex. 2. What will 6075 lbs. of coal come to at 77 1-2 cents per hundred? Ans. 47.08.

```
    6075
    ,775=77 1-2
 ---------
 47.08.125
```

We point off from the product 3 figures, then two more for cents; those at the left hand will be dollars.

Ex. 3. What will 9 car loads of coal cost, weighing as follows, at $1.14 per hundred? Ans. $230.73.

```
     3630
     2210
     3700
     1310
     1830
     2010
     1810
     1850
     1890
    -----
    20240
     1.14
 ---------
 230.73.60
```

Ex. 4. What will the freight on 8 boxes of sugar amount to, weighing 1685 lbs. at 55 cts. per hundred? Ans. $9.25.

```
 1685
  .55
-----
9.26.75
```

Ex. 5. If meal is $1.50 per hundred, how much can be bought for $75.45?

```
150 : 7545 :: 100
             100
       ------
   150)754500
       ------
        5,030 ans.
```

Ex. 6. How many tons are there? Ans. 2 103-200.

```
2,000)5030
      ------
     2 103-200
```

Ex. 7. If meal is $1.60 per hundred, how much can be bought for 10 cents? Ans. 6 1-4 lbs.

```
160 : 10 cts. :: 100 lbs.
                  10
                 ---
            160)1000(6
                 960
                 ---
                  40
                  16 oz. pound.
                 ---
             160)640
                 ---
                   4
```

GOODS BOUGHT AND SOLD BY THE THOUSAND.

Rule.—Multiply the number of feet or bricks, as the case may be, by the price, and point off from the product the same manner as the rule on p. 163.

Ex. 1. What will 822 feet of boards amount to at $7.50 per thousand? Ans. $6.16.

```
  822
  750
-------
6,16.500
```

Ex. 3. What will 28 feet boards come to at $3.75 per thousand? Ans. 10 cents.

```
   375
    28
-------
,10,500
```

MISCELLANEOUS EXAMPLES.

Gold at Premium.—You will have no difficulty in determining the discount on greenbacks to correspond with the premium on gold, if you will fix clearly in your mind the fact that discount is always to be ascertained by dividing the amount in greenbacks by the price in gold.

For example, 1st—when gold is worth 25 per cent. premium, so that a hundred dollars in gold is worth $125 in greenbacks, you have only to divide $100 by 100, increased by the rate per cent., and you get $80, which is the value in gold of the $100 in greenbacks, and this gives a discount of 25 per cent. on greenbacks when gold is at 25 per cent. premium?

Add 25 per cent. makes						
100	per cent. is written thus,					1.00
25	"	"	"	"		.25
125	"	and	"	"		1.25

1.25)100.00

$80 ans.

Ex. 2. If you reverse the process and reckon up what $80 in gold is worth, 25 per cent. premium, you will find that it comes to just $100 in greenbacks.

125 per cent. is written thus, 1.25
80,00

100.0000

Ex. 3. If gold is 25 per cent. premium, and I have $100 in gold, how many dollars in greenbacks shall I get for it? Ans. $125.

1.25
100.00

$125.00.00

Ex. 4. If gold is at 25 per cent. premium, what is $1 greenback worth? Ans. 80 cts.

1,25)1,00.00

0,80 ans.

The decimal places in the quotient exceed those in the divisor by 2, counting the ciphers annexed, we point off two decimal places in the quotient for cents.

Ex. 5. If flour is sold for $9 per barrel in greenbacks, and gold is 52 per cent. premium, what must it be sold for in gold to bring its actual value?

100 per cent. is written thus, 1.00
52 " " " " .52

152 " and " " 1.52

Add 52 per cent. makes

1.52)9.00

$5.92 ans.

Ex. 6. If government bonds are 12 per cent. premium, what must I pay for a bill of $5786; what would be the amount at 2 per cent. discount?

Ans to the 1st; 6480.32.
" " 2d, 5670.28.

5786.00
112 per cent. is written thus, 1.12

$6480.32.00

5786.00
98 per cent. is written thus, .98

5670,28,00

Point off from the product the same as Ex. 8, p. 3.

Ex. 7. France once owed the United States 18,000,-000 francs; what is the amount in federal money?

18.000,000
.186=1 franc.

3,348,000,000

There are 3 decimal places in one factor; hence we point off from the product 3 decimal places; all at the left will be dollars—making three millions, three hundred and forty-eight thousand dollars.

Ex. 8. There are four concrete sidewalks, the first is 120 feet long, 5 1-2 feet wide. The second is 12 feet long and 4 feet wide, and 2 more, each 19 feet long and 4 feet wide; what will they cost at 80 cents per square yard?

```
              120              12
               5.5=5 1-2        4
              ----             --
              6600             48 sq. ft.
               48
              152
              --                            19
9 ft. sq. yd.)860                             4
              -----                         --
               95.55                        76
              ,80                            2 walks.
              ------                        --
              $76.44.00                     152
```

Ex. 9. If tea is worth 1.75 per pound, how much can be bought for 40 cents. Ans. 3 oz. 10 dr.

Statement.

```
175 : 40 :: 16 oz. a pound.
            40
            --
      175)640(3
          525
          ---
          115
           16 drams an ounce.
          ----
          1840(10
          1750
```

Ex. 10. There is a street 40 rods long and 4 rods wide; how many square yards does it contain? How many squares?

```
        40
         4
        --
       160
        30 1-4 square yards 1 square rod.
       -----
 1.00) 48.40 sq. yds ans. to the 1st.
       -----
       48.40=48 2-5 squares, ans. to the 2d.
```

Ex. 11. If Brian Hill coal is selling for $8.50 per ton, how much can be bought for $20?

850 :2000 :: 2000
2000

850)4000000

4705.8 ans.

See rule in Simple Proportion, p. 98.

Ex. 12 How many cubic feet in a box 4 1-2 feet long, 2 1-2 feet wide and 2 feet deep? Ans. 22 1-2 ft.

4.5=4 1-2
2.5= 1-2

11.25
2

22.5)

Ex. 13. How many superficial feet are there in a piece of marble 5 feet 5 inches long and 3 feet 4 inches wide? Ans. 18 ft. 8 seconds.

5	5	
3	4′	
1	9′	8″
16	3′	
18	0′	8″

Ex. 14. How many cubic feet, and how many feet, board measure, are there in a plank 2 feet 11 inches long, 1 foot 6 inches wide and 5 inches thick?

2	11′			
1	6′			
1	5′	6″		
2	11′			
4	4′	6″		
	5′			
1	9′	10″	6‴	ans. to the 1st.
		4	4′	6″
				5 inches thick.
		21	10′	6″ ans. to the 2d

Ex. 15. How many feet of wood, at $4 per cord, shall be given for 50 cents?

```
   400 : 50 :: 128 ft.
                50
               -----
         4.00) 64.00
               -----
                 16 ft. ans.
```

Ex. 16. How many tiles, 9 inches square, will cover a hall 126×16 feet?

```
 9                      126
 9                       16
--                      ----
81 sq. inches.          2016
                         144 in. sq. ft.
                    ---------
                81) 290304
                    ---------
                        3584 tiles ans.
```

Ex. 17. If a pole 16 feet long cast a shadow 22 feet, what is the height of a steeple whose shadow is 216 feet?

```
22 : 216 :: 16
      16
     ----
22)3456
   --------
     157 1-11 ans.
```

Ex. 18. How long must I keep $300 to balance the use of $500 which I lent a friend 4 months?

```
     300 : 500 :: 4
                   4
             ------
       3.00) 20.00
             ------
              6 2-3 mos. ans.
```

Ex. 19. How many times will a hind wheel of a

carriage 8 feet 6 inches in circumference, turn around 6 miles, 3 furlongs and 20 rods?

```
          320 rods a mile.
            6 miles.
         ----
         1920
          120 rods=3 furlongs.
           20 rods.
         ----
         2060
          16.5=16 1-2 ft. a rod.
         -------
   8.5) 33990.0
        -------
        3998.82 ans.
```

Ex. 20. How many pounds of sugar shall be given for 13 cents per pound for $1?

Ans. 7 lbs., 11 oz. 1 d.

```
13)100(7
    91
    --
     9
    16 oz. pound.
    --
   144(11
    13
    --
    14
    13
    --
     1
    16 drams an ounce.
    --
    16(1
    13
    --
```